BURLEIGH DODDS SCIENCE: INSTANT INSIGHTS

NUMBER 45

Ensuring animal welfare during transport and slaughter

Published by Burleigh Dodds Science Publishing Limited
82 High Street, Sawston, Cambridge CB22 3HJ, UK
www.bdspublishing.com

Burleigh Dodds Science Publishing, 1518 Walnut Street, Suite 900, Philadelphia, PA 19102-3406, USA

First published 2021 by Burleigh Dodds Science Publishing Limited
© Burleigh Dodds Science Publishing, 2021, except the following: The contribution of Luigi Faucitano in Chapter 3 is © Her Majesty the Queen in Right of Canada. All rights reserved.

British Library Cataloguing in Publication Data
A catalogue record for this book is available from the British Library

ISBN 978-1-80146-223-5 (Print)
ISBN 978-1-80146-224-2 (ePub)

DOI 10.19103/9781801462242

Typeset by Deanta Global Publishing Services, Dublin, Ireland

Contents

Series list

Title	Series number
Sweetpotato	01
Fusarium in cereals	02
Vertical farming in horticulture	03
Nutraceuticals in fruit and vegetables	04
Climate change, insect pests and invasive species	05
Metabolic disorders in dairy cattle	06
Mastitis in dairy cattle	07
Heat stress in dairy cattle	08
African swine fever	09
Pesticide residues in agriculture	10
Fruit losses and waste	11
Improving crop nutrient use efficiency	12
Antibiotics in poultry production	13
Bone health in poultry	14
Feather-pecking in poultry	15
Environmental impact of livestock production	16
Pre- and probiotics in pig nutrition	17
Improving piglet welfare	18
Crop biofortification	19
Crop rotations	20
Cover crops	21
Plant growth-promoting rhizobacteria	22
Arbuscular mycorrhizal fungi	23
Nematode pests in agriculture	24
Drought-resistant crops	25
Advances in crop disease detection and decision support systems	26
Mycotoxin detection and control	27
Mite pests in agriculture	28
Supporting cereal production in sub-Saharan Africa	29
Lameness in dairy cattle	30
Infertility/reproductive disorders in dairy cattle	31
Antibiotics in pig production	32
Integrated crop–livestock systems	33
Genetic modification of crops	34

Chapter 1

Optimizing welfare in transport and slaughter of cattle

Jan Shearer, Iowa State University, USA

1 Introduction

Safeguarding the welfare of animals during transport and at slaughter requires an understanding of the risks encountered during each phase of the journey and at the destination. Pre-transport assessment of an animal's fitness for travel can reduce problems during transport and at slaughter. If the trip or time in transit is likely to be lengthy, it is important that animals are well fed and hydrated prior to departure. During transport, the welfare of cattle is affected by the experience of the driver, time and distance travelled and weather conditions. The scarcity of slaughter facilities throughout the United States has resulted in longer travel times and distances, contributing to driver fatigue and a greater potential for livestock hauler accidents. The welfare of cattle at slaughter is affected by the condition of animals on arrival and opportunities for them to rehydrate and rest in lairage. Proper scheduling should prevent the need for animals to remain in lairage for extended periods prior to slaughter.

http://dx.doi.org/10.19103/AS.2020.0084.13

Cattle handling, movement of animals to the stunning box, stunning accuracy and slaughter procedures, also affect the welfare of cattle at abattoirs.

To provide context for the development of strategies to optimize the welfare of cattle during the pre-transport, transport and slaughter phases, it is important to understand why animals are removed from herds; what are the common modes of transport; and, what are some of the current observations regarding the condition of animals at slaughter? Not all cattle are culled for reasons of disease or injury. Most animals culled from herds are healthy but must be removed because of inferior performance relative to an available herd replacement. Dairy cows leave herds on average at a little less than 5 years of age, whereas the longevity of beef cows is approximately twice that of dairy cows at 10 or more years.

Modes of transport for healthy cattle culled from farms and ranches to markets and slaughter are generally via semi-pulled potbelly and light truck gooseneck or bumper-pull type trailers. Compromised cattle should be hauled to the nearest possible slaughter facility and in trailers where they can be isolated from other animals. Muscling and body condition of cull dairy cows arriving at slaughter plants is consistently low. This not only reduces carcass value but increases susceptibility to bruising during transport.

Although the incidence of obvious physical defects in cattle at slaughter is low, more than one-third of hides are damaged by the practice of branding. Hide-off evaluation of carcasses at slaughter reveals that roughly two-thirds of cattle carcasses require some degree of trimming because of bruising. The incidence of 'dark cutters' (i.e. carcasses that are dark in color) is associated with pre-slaughter stress. Branding, bruising and dark cutting beef are important indicators of poor welfare. They are not only very costly problems, but they are also largely preventable.

2 Culling and permanent herd removals

2.1 Voluntary versus involuntary culling

Culling is an animal removal strategy intended to improve productivity and profitability in dairy and beef cattle herds. Reasons for culling are often described in terms of 'voluntary' or 'involuntary'. Cows that are removed from herds for voluntary reasons are those removed at the will of the owner. Involuntary reasons for removing an animal from the herd are those against the will or conscious control of the owner. Dairy cows that leave herds for voluntary reasons are primarily influenced by milk price, market or slaughter value of cows, cost and availability of replacements and housing capacity. Involuntary herd removals are generally for health-related conditions that reduce an animal's performance below cost-benefit breakeven levels. Depending upon the specific condition, these animals present the greatest welfare concern in the context of transport to markets and slaughter.

2.2 Surplus replacements and bull sales

In situations where the dairy has an over-supply of replacements, it may elect to sell animals for 'dairy purposes', that is, sell animals to other dairies for making milk. Some dairies raise bulls for sale to herds that rely on natural service for some if not all of their cattle breeding needs. Purebred beef herds with an excess of replacements may sell heifers, cows or bulls to other herds seeking genetic improvement. Sales such as these do not fit the cull animal criterion. These animals are not removed for reasons of poor performance; rather, they leave because the herd has animals in excess of its needs or is in the business of marketing superior genetics.

2.3 Culling based upon performance

The primary product of dairy farming is milk; in the cow/calf industry, the product is a calf on an annual basis. Calves from cow/calf operations may go directly to a feedlot, whereas others may make an intermediate stop at a backgrounding operation, where they are readied for movement to the final production phase in a feedlot. The feedlot's objective is to produce a finished steer or heifer ready for slaughter.

Performance and production efficiency are key factors in the profitability of each enterprise. A dairy cow that fails to produce milk at a level above the cost of maintaining her becomes a cull and is moved to slaughter. Similar decisions are made for a beef cow that fails to become pregnant or calves in the stocker or feedlot industries that fail to grow or gain at profitable levels. Bulls are valuable as long as they are fertile and physically capable of producing pregnancies in cows. When their capacity for producing pregnancies in cows is lost, they too are moved onto slaughter. This is the truest sense of the term 'cull', that is, to remove an animal from a herd that is less productive so that a more profitable animal may replace it.

2.4 Removal (i.e. cull) rates for dairy and beef herds

The annual herd removal rate for dairy cattle operations in the United States is approximately 30% (Geiger, 2016). Rates vary considerably and are influenced by multiple factors including parity, milk production, fertility and cow health. Some herds may temporarily reduce or adjust cull rates when planning for herd expansion. Another driver of cull rates is the availability of herd replacements. The advent of sexed semen has created a surplus of dairy heifer replacements allowing for more stringent culling of older, less profitable cows. De Vries calculated that the normal productive life of a dairy cow in the United States was 2.63 years or 31.6 months. Assuming an average age at first calving of 25.5 months, plus a cow's average productive life of 31.6 months, the average dairy cow remains in the herd for approximately 57.1 months or 4.8 years of age before being removed from the herd (De Vries, 2013).

Annual cull rates for cow/calf herds are also variable, but estimated to be in the range of 10–20%. A beef cow must successfully deliver and raise a calf each year, and in addition be healthy and fertile enough to become pregnant. Failure to accomplish these feats makes her a candidate for culling from the herd. Over half of the cows (52.6%) sold from herds in 2007 were 10 years old or older than that. According to USDA statistics, 41.8% of beef cattle operations reported culling cows that did not become pregnant. Other reasons for culling cows included: age and bad teeth 55.7%, temperament 16.6%, other (unspecified) reproductive health problem 13.4%, economics (drought, need to reduce herd size, and market conditions) 10.9%, producing calves 10.7%, physical unsoundness 9.6%, udder problems 9.2% and eyesight problems 7.1% (USDA, 2007–08, https://www.aphis.usda.gov/animal_health/nahms/bee fcowcalf/downloads/beef0708/Beef0708_dr_PartIV_1.pdf). The removal of less-productive cows creates opportunities for herds with an ample number of healthy productive replacements.

3 Transport of cattle to markets and slaughter facilities

A rough estimate of the number of cattle transported to dairies, growing operations, livestock markets and/or slaughter plants likely exceeds 400 000 each day. The overwhelming majority are hauled in semis pulling potbelly trailers and pick-ups with a gooseneck or bumper-pull trailer in tow. In cattle states throughout the United States, it is virtually impossible to avoid encountering a livestock hauler transporting cattle or other livestock to some destination whether it be a market, packing plant or just from one farm to another.

3.1 Potbelly trailers

Feeder calves, market and slaughter cattle are often hauled in semi-pulled potbelly or modified potbelly trailers. The term potbelly refers to the 'drop down belly' design of the trailer that when loaded increases stability (reducing the potential for rollover accidents) and permits better space utilization. These trailers have an upper deck and cut gates that permit proper distribution of weight and cattle into the upper (top) and lower (bottom) nose, upper deck, belly, doghouse (also known as the jail), which is the upper compartment in the back of the trailer. Depending upon the number of axles some of these trailers can haul nearly 90 000 lb (40 823 kg).

Potbelly trailers can be modified into a fat cattle/feeder cattle combo trailer with the versatility to haul both feeder calves and finished beef cattle (Fig. 1). Fat trailers are designed to haul finished feedlot steers and heifers only (Fig. 2). The differences between these types of trailers include the presence

Configuration for a Fat-Feeder Combination Trailer

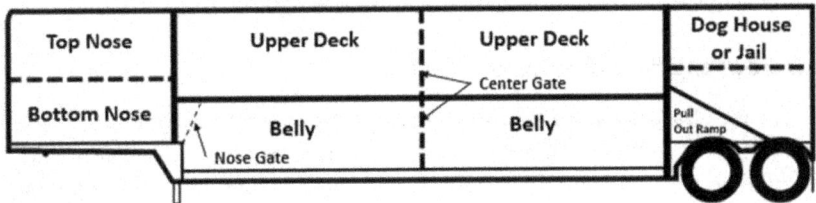

Figure 1 Configuration for a trailer designed to haul fat cattle and/or feeder steers and heifers (adapted from Master Cattle Transporter Guide, https://www.bqa.org/Media/BQA/Docs/master_cattle_transporter_guide-digital.pdf).

Configuration for a Fat Cattle Trailer

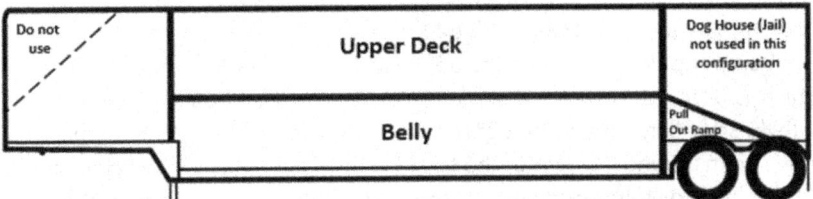

Figure 2 Configuration for a trailer designed to haul fat cattle only (adapted from Master Cattle Transporter Guide, https://www.bqa.org/Media/BQA/Docs/master_cattle_transporter_guide-digital.pdf).

or absence of the 'jail' or 'doghouse' in the upper rear compartment of the trailer. This compartment (the doghouse) is present in the fat/feeder combo configuration. Both configurations have a slide-in or fold-up ramp that leads to the upper deck above the belly. One feature of note is that the clearance height of the entrance into the 'belly', or lower compartment of the trailer is approximately 2–3 inches (5–8 cm) shorter in the fat/feeder combination trailers. This has important implications for taller cattle breeds such as Holsteins, which are more likely to injure themselves on entry and exit of from the belly of the trailer (Master Cattle Transporter Guide, National Beef Quality Assurance Guide for Cattle Transporters, https://www.bqa.org/Media/BQA/Docs/master_cattle _transporter_guide-digital.pdf).

In order to assess the number of traumatic events cattle may encounter in the process of unloading, researchers counted the number of times cattle hit any part of the trailer during the off-loading process at packing plants. A total of 9860 animals in 75 lots were observed at three slaughter plants being off-loaded from 275 trailers. The prevalence of traumatic events was highest for Holstein cattle hauled in fat/feeder combination trailers. Researchers also found a significantly higher prevalence of carcass bruising in Holstein cattle (76.6%) compared to 66.6% in beef breed cattle (Lee et al., 2017). The most likely reasons are the larger frame and height of Holsteins, which would naturally make them more susceptible to trauma and bruising.

3.2 Gooseneck and bumper-pull trailers

Gooseneck trailers have a long neck that reaches over the tailgate of the truck box to fit over a ball hitch in the truck bed. Livestock trailers are available in lengths ranging from 16 ft (4.9 m) to 36 ft (11 m). Trailer floors are either wood or in some cases aluminum. Depending upon the length of the trailer most will have at least one cut gate. An 8 ft (2.4 m) × 24 ft (7.3 m) trailer should accommodate 3–5 1200–1400 lb (544-635 kg) animals per compartment. Approximately 20–25% of the weight of the load should be on the hitch. This adds stability when the trailer is loaded because the weight is better distributed over the axles and truck chassis (Livestock trailer safety, 2012, http://www.exte nsion.org/pages/64391/livestock-trailer-safety).

Bumper pull trailers attach to a ball hitch on the back bumper of the truck, which is connected to the truck's frame. Weight distribution is more critical with these trailers. Generally, 60% of the weight should be in front of the trailer axle(s) with around 10–15% of the total weight on the hitch. Insufficient weight on the hitch or tongue relative to the rear of the trailer can cause the trailer to swing the rear of the truck from side to side making it extremely difficult to control the vehicle (Livestock trailer safety, http://www.extension.org/pages /64391/livestock-trailer-safety) .

Truck and trailer maintenance (proper tire inflation and wheel bearing maintenance) is extremely important. Breakdowns with cattle onboard in hot and humid or extremely cold weather can be very detrimental (if not deadly) for the welfare of cattle on board. For trailers with wood floors, it is important to remember that wood floors eventually rot. If an animal happens to place a foot on a weak spot in the floor and it breaks through, it can have disastrous consequences (Stalsberg et al., https://fyi.extension.wisc.edu/wbic/files/2018 /03/3T-handout-1done.pdf).

3.3 Transporting compromised cattle

Cull cows compromised by lameness or other condition, yet fit for human consumption and short-distance travel, should be transported to the nearest possible slaughter plant. They should not be transported to markets where there is the potential for long distance travel on the next leg of their journey. If several animals are being transported, the animals with the best mobility should be loaded first and toward the front of the trailer. Lame or weaker animals should be loaded last and kept in a separate section from other animals. This helps avoid further injury that may occur in transit from bullying by other animals or trampling should the animal fall.

Cull cows sold at markets are often hauled on potbelly trailers to their final destination at a packing plant. The hauler must carefully evaluate the size and fitness of these animals, particularly when long-distance travel may be required. Weak or lame animals may have difficulty walking up and down ramps upon entry and exit from trailers. In potbelly trailers, cattle must be able to walk up and down ramps into the belly of the trailer as well as to the upper deck. They must also adjust to co-mingling with other animals and vehicular movement during transport. For many reasons, the welfare of cull cows and particularly those with special needs requires additional care and handling (Cockram, 2019).

4 Trends in transport of dairy and beef cattle

NAHMS data for all operations studied found that 92% of dairies sold cows through livestock markets, auctions or stockyards. Thirty-seven percent of dairy operations sent cull cows directly to slaughter. Smaller operations tended to send cull cows to markets rather than direct to slaughter plants. As herd size increased, the average number of cows shipped to a market, auction or stockyard, or directly to a packer increased (USDA, 2018).

Cattle leaving herds by the number of lactations was 22.2% for first, 51.9% for second to fourth, and 25.9% for fifth or higher lactations. Of cows permanently removed, 9.5% were sold as replacements and 21.1% were sold for low milk production. Cows that left herds for involuntary reasons included 21.2% for infertility, 16.5% for clinical mastitis and 7.2% for lameness. By stage of lactation, researchers observed that approximately 20% of cows left herds in early lactation, 24% at mid-lactation and nearly 50% (49.3%) of cows were removed in late lactation. Cows with serious health conditions and low milk production generally left herds early in lactation (USDA, 2018, https://www.aph is.usda.gov/animal_health/nahms/dairy/downloads/dairy14/Dairy14_dr_Par tIII.pdf).

In terms of distance travelled to a market, auction or stockyard, 14.3% of shipments were within 1 (1.6 km) to 9 (14.5 km) miles and 63.4% of shipments were within 10 (16 km) to 49 (79 km) miles. This means that (14.3 % + 63.4%) roughly, 80% of cattle travel less than 50 (80.5 km) miles to a market, which sounds reasonable until one considers that this is likely only the first stop on their journey. From here, animals will travel to a slaughter plant, which for cattle in certain parts of the West or Southeast could mean a very long trip on the second leg of their journey (Edwards-Callaway et al., 2019). NAHMS data indicates that 30% of dairies send cows to slaughter plants across state lines (USDA, 2018, https://www.aphis.usda.gov/animal_health/nahms/dairy/do wnloads/dairy14/Dairy14_dr_PartIII.pdf). However, depending upon the needs of the cattle buyer, cattle may be transported much further (Edwards-Callaway et al., 2019).

For shipments going directly to slaughter, 7.7% travelled from 1 (1.6 km) to 9 (14.5 km) miles, 42.8% from 10 (16 km) to 49 miles (79 km), 38.3% from 50 (80.5 km) to 249 (400 km) miles and slightly over 11.2% travelled more than 250 (402 km) miles. By simple calculation, approximately 50% of animals travel less than 50 miles (80.5 km), but more than 11% travel in excess of 250 miles (402 km). Any animal that exhibits weakness or immobility, yet can still be safely processed for human use, should be taken to the closest possible slaughter plant (USDA, 2018, https://www.aphis.usda.gov/animal_health/nahms/dairy/downloads/dairy14/Dairy14_dr_PartIII.pdf).

An excellent review of the current challenges to the welfare of cattle during transport to markets and slaughter in the United States was published by Edwards-Callaway (Edwards-Callaway et al., 2019).

4.1 Transport of beef cattle

According to the recent National Beef Quality Audit, cattle were found to be in transit for a mean of 6.7 h and travelled a mean distance of 283 miles (455.7 km). Three loads travelled more than 24 h and travelled distances greater than 994 miles (1600 km) (Harris et al., 2017). Long-distance travel is associated with an increased amount of carcass bruising, shrink (dehydration) and stress. Longer trips and transit times (i.e. beyond 24 h) also require stops in route to provide feed and water in accordance with the Beef Quality Assurance program guidelines. The maximum transportation times collected were 12 h for fed cattle and 39 h for cows/bulls. The longer and farther animals are transported the greater the impact on carcass quality traits (bruising and dark cutting beef) at slaughter.

The average number of cattle per load was 26, although load size ranged from 1 to 47 animals per trailer (Harris et al., 2017). Recommended minimum space requirements for cattle weighing 360 kg (800 lb), 454 kg (1000 lb), 545 kg (1200 lb) and 635 kg (1400 lb) are .97 m^2 (10.4 ft^2), 1.11 m^2 (12 ft^2), 1.35 m^2 (14.5 ft^2) and 1.67 m^2 (18 ft^2), respectively. Range in space allowance for cattle with horns is an additional .05 m^2 to .09 m^2 (0.5-1 ft^2) (Grandin, 2013). The average space allotment per animal was 2.3 m^2 (24.8 ft^2) for all loads studied indicating that recommendations for space per animal were met in the loads observed (Harris et al., 2017). Additional guidelines on trailer compartment dimensions and weight of the cattle hauled may be found in the Master Cattle Transporter Guide at: https://www.bqa.org/Media/BQA/Docs/master_cattle_transporter_guide-digital.pdf.

Trailer dimensions varied, as there were several types of trailers used. Pot belly trailers were the primary type of trailer used by 67.1% of haulers to transport cows and bulls to market, followed by gooseneck trailers at 30.3%. On average, haulers using pot belly trailers (n=95) used five compartments for separating cattle. Separating cattle using a center gate and compartments reduces carcass bruising and animal welfare concerns. It might seem that separating cows and heifers from bulls during transport might reduce the incidence of bruising; however, 64.4% of loads hauling both cows and bulls did not separate them according to sex (Harris et al., 2017).

5 Muscling and body condition of cull cows arriving at slaughter plants

Anecdotal information and findings of the 2016 National Beef Quality Audit (NBQA) indicate that a significant number of cull cows from the dairy industry are arriving at slaughter plants in less than satisfactory condition. As an example, the recent NBQA scored cattle for muscling as follows: score 1 = extremely light muscled to 5 = extremely heavy muscled. Statistics indicate that 93.8% of dairy cows were extremely light muscled or light muscled (score of 1 or 2); only 4.2% scored 3, 1.9% scored 4 and none scored 5. By comparison, only 34.7% of beef cows had extremely light or light muscling scores. Cattle with the greatest frequency of a muscle score of 3 were beef cows, beef bulls and dairy bulls (Harris et al., 2017).

Similar observations were made for body condition. Body condition scoring (BCS) is a useful management tool for determining nutritional needs and body reserves in cows. A 9-point scoring system is used for beef cows and beef bulls, whereas dairy cows and dairy bulls are scored using a 5-point scoring system with half scores. For perspective, a beef cow with a score of 5 or 6 is considered to be in ideal condition. A BCS of 1 is considered to be extremely thin and score of 9 would be considered a very obese cow. Using the 5-point scale with half scores for dairy cows and bulls provides a 9-point system similar to that for beef cattle. A cow with a BCS of 1 is considered a very thin cow and one with a BCS of 5 is considered very obese. Regardless of the stage of lactation, a cow with a BCS of 2 would be on the edge of being too thin and one with a BCS of 4 would be too fat. Lactating dairy cows scoring between 2.5 and 3.5 are considered to be within the ideal range.

The mean BCS recorded in the 2016 NBQA for beef cows (n = 1 911) and beef bulls (n = 406) was 4.7, whereas the mean BCS for dairy cows (n = 2 878) and dairy bulls (n = 121) was 2.6 and 3.3, respectively. Compared with data collected from 2007 (Nicholson, 2008) body condition of cows (both beef and dairy) improved from 2007 to 2016. Nonetheless, 9.3% of dairy cows and 0.8% of dairy bulls had BCS characterized as 'too low' (i.e. from 1.0 to 1.5). For the sake of comparison, 7.6% of beef cows and 5.7% of beef bulls had BCS considered too low (Harris et al., 2017).

The message to beef producers and dairy farmers is that when or if economically feasible, producers should strive to improve body condition and muscling in cattle determined to be light prior to marketing and/or slaughter. In some cases, marketing cattle while still in good condition, or increasing feed for light-muscled cattle and those with lower BCS has been shown to increase muscle and fat in thin under-conditioned animals (Matulis et al., 1987; Schnell et al., 1997). Improvements in body condition and muscling are also key determinants in market readiness, fitness for travel and welfare of cattle.

5.1 Physical defects observed in beef and dairy culls at slaughter

Based upon observations from the 2016 NBQA, the majority of cattle observed at slaughter facilities had no obvious injuries or physical defects. This suggests that most cattle were culled for reasons such as reproductive failure, undesirable behavioral characteristics or other non-physically obvious reasons. Physical defects that were observed in beef cows (n= 1 912), beef bulls (n = 402), dairy cows (n = 2 855) and dairy bulls (n = 120) included the following: bottle teats, broken penis, failed suspensory ligament, foot abnormality, full bag, lumpy jaw, mastitis, multiple udder problems, retained placenta, swollen joints and warts. At least one of these visible defects was present in 44.1%, 32.1%, 27.9% and 24.1% of dairy cows, beef bulls, beef cows and dairy bulls, respectively. Foot abnormalities (12.7%), swollen joints (4.7%) and the presence of an abscess (2.7%) were observed with the greatest frequency in beef bulls. The percentage of foot abnormalities in beef cows was 3.4% followed by 2.5% in dairy cows (Harris et al., 2017).

Chilean researchers recently reported on the pre- and post-transport effects on the welfare of 237 cull cows shipped directly from farms with an average journey length of 5 h 22 min. The shortest journey was 45 min and the longest was a maritime-terrestrial transport from a farm in the Magallanes Region that lasted 46 h 55 min. Cattle on the longest journey received neither water nor food at any time during the trip (Sánchez-Hidalgo et al., 2019). All cattle had at least one visible health problem. Fifty-two percent (124/237) of cattle had low body condition, 50% (119/237) mammary gland disorders and 24% (n = 57) were lame (Sánchez-Hidalgo et al., 2019).

6 Hide and carcass bruising evaluation

Branding reduces the value of cattle hides. It also causes acute as well as persistent pain (Tucker et al., 2014). Cattle that received a brand experienced pronounced pain in the days immediately following branding that lasted for 8 weeks following the date of procedure. Of cattle surveyed in the NBQA study, 22.7% had a least one visible brand. The percentage of beef cattle hides with brands was 35.7% compared to dairy cattle at 10.7%. Although hot-iron branding is the most permanent form of identification, it devalues the hide by as much as $7.47 USD per animal (National Cattlemen's Beef Association, 2017) and presents a significant animal welfare concern. Better methods (less painful) of animal identification are needed.

6.1 Hide-off evaluation of carcass bruising

Carcass bruising may occur during any phase of the pre-transport, transport or post-transport process. Bruises require trimming that costs the cattle industry millions of dollars annually. In the 2016 NBQA, 35.9% of cattle (n = 4 262) had carcasses with no bruises, 67.3% bruises of minimal severity, 45.1% characterized as major, 4.9% critical and 1.4% as extreme. Observations in bulls (n = 389) revealed 57.5% with no bruises, 42.4% with minimal, and 21.9% with major bruises. When bruises are minimal in severity, they require minimal amounts (less than 0.45 kg) of surface trim. Comparing dairy with beef cattle, multiple bruises were observed in carcasses of 41.3% of dairy and 24.0% of beef cow carcasses. Similarly, 25.2% of dairy bull and 13.5% of beef bull carcasses had multiple bruises (Harris et al., 2017). At least part of the solution for reducing the risk of bruising is to improve body condition of cows and bulls.

In cow carcasses, the greatest percentage of bruises were located on the round and sirloin, whereas in bulls bruises occurred with greater frequency on the brisket, plate (behind the brisket and below the rib), and flank region when compared to cow carcasses. Bruises occurring within 24 h of harvest are usually a direct result of handling practices and facility design (Harris et al., 2017). Cows with poor body condition also had more severe bruising.

6.2 Predisposing factors and causes of bruising

Bruises are an indicator of poor animal welfare and an important cause of economic loss to packers (Grandin, n.d.; Huertas et al., 2010; Lee et al., 2017). The literature is replete with studies that demonstrate that bruising may occur at the farm, during transport and at slaughter (Harris et al., 2017; Grandin, n.d.; Jarvis et al., 1995; Sánchez-Hidalgo et al., 2019; Bethancourt-Garcia et al., 2019). It is important to identify the cause of bruises and at what point of the transport phase they may have occurred.

Pre-slaughter factors include the degree of muscling and BCS, gender and breed, horned versus polled cattle, facilities and handling at the farm of origin, and use of prods (Harris et al., 2017; Jarvis et al., 1995; Sánchez-Hidalgo et al., 2019; Bethancourt-Garcia et al., 2019). In the recent study by Sánchez-Hidalgo et al., poor body condition increased the likelihood of bruising by 20-fold.

Bruises occurring during transport may be a consequence of size and type of transport vehicle (i.e. larger versus smaller trucks) (Lee et al., 2017), animal density, the distance travelled and whether animals have been initially transported to markets or directly to slaughter. Inexperienced drivers or those who are prone to driving and taking corners too fast, making sudden stops or accelerating too rapidly may be to blame for an increase in bruising. When bruising is extensive, it may indicate that the animal had fallen and was trampled in transit (Tarrant et al., 1988; Grandin, https://www.grandin.com/references/cause.bruising.html). Holstein cattle experienced more traumatic events than beef breeds when hauled in potbelly trailers normally used for fat cattle and feeder calves. Researchers suggested that the larger frame scores/sizes were likely the predominant contributing factor (Lee et al., 2017).

If bruises are packing plant-related, carcasses will have bruises in the same anatomic location regardless of the origin of the cattle (Grandin, https://www.grandin.com/references/cause.bruising.html). Plants should inspect unloading ramps, alleyways, chutes and restrainers for evidence of broken parts or protruding objects that might cause bruising or injury. They must insure that employees understand proper animal handling techniques and places in the processing system where the potential for bruising may occur. Understanding the causes of bruising at the farm or feedlot, in transit or at the packing plant isthe first step in reducing the incidence of bruising and improving the welfare of cattle during transport and slaughter.

6.3 Dark cutting beef (DCF)

'Dark cutters' are a major meat-quality concern, but they are also indicative of welfare issues. It occurs in cattle that undergo a significant degree of pre-slaughter stress such as that associated with transport or rough handling at some point during the pre-slaughter process. Dark cutting beef (DCB) is described as dark red in color, lacking desirable tenderness, having an objectionable flavor and a short shelf-life. The dark color of the meat is caused by a depletion of glycogen stores in muscle tissues that raises the pH thereby facilitating bacteria growth on the meat's surface. In the absence of glycogen, the bacteria use glucose and amino acids as substrates for their metabolism and in the process generate putrefactive amines that give DCB its characteristic spoiled odor and objectionable taste.

An extensive review by Ponnampalam et al. (2017) indicates that on-farm contributing factors may include farm and feedlot feeding and management practices, animal genetics, breed and many others. Off-farm factors include marketing and transport conditions, lairage in route to and at the packing plant, and stunning and slaughter practices at the packing plant (Ponnampalam et al., 2017). It is beyond the scope of this chapter to discuss this issue in detail; instead, readers are directed to the review by Ponnampalam et al. for additional information on the subject of DCB.

7 Optimising welfare during the pre-transport stage

Transport is one the most stressful events cattle ever experience. It is compounded by the presence of disease or injury, recent weaning, co-mingling with unfamiliar animals, exposure to multiple handlers, inclement weather conditions, long-distance travel without feed or water and arrival to a completely new environment. Pre-planning at the pre-transport and transport phases along with proper handling and management of cattle at slaughter can minimize the negative impacts on end-product quality and animal welfare.

7.1 Pre-transport

The pre-transport phase includes decision-making on animals to be culled or removed from the farm or ranch. Once this is determined, the next assessment is the animal's fitness for transport be it to a livestock market or to slaughter. Animals that are strong, healthy and free of conditions affecting their mobility may be considered for sale at a livestock market or auction. Cattle with physical limitations (mild or moderate lameness, for example) yet still fit the requirements for human consumption should be sent to the nearest slaughter facility. Other procedures that may appreciably affect the welfare of cattle during the pre-transport phase are penning and handling at loading.

7.2 Determining fitness for travel

'It is impossible to assure good animal welfare during transport if the animal is unfit' (Cockram, 2019). Animals sent for slaughter with preexisting conditions are more likely to die in transit, become non-ambulatory, or require euthanasia on arrival to the plant (Cockram, 2019). In Canada, compromised cattle may be transported if they are sent directly to a nearby slaughter facility. They must be carefully loaded and segregated to avoid possible injury from interaction with other animals. They must be loaded last and unloaded first. Producers must weigh the potential for a financial return with the risks of animal suffering. They must consider the possibility of financial loss from mortality or condemnation of the carcass and the likelihood that an inspector or veterinarian may decide that the animal should not have been sent (Cockram, 2019).

The importance of carefully examining compromised animals prior to loading and movement to slaughter cannot be overemphasized. The Canadian Food Inspection Agency conducted a survey in 2001 to determine the number of non-ambulatory cattle arriving at federally inspected slaughter plants and auction markets in Canada. Of the 7382 non-ambulatory cattle arriving at these sites, 89.8% were dairy cattle with only a little over 10% identified as beef cattle (Doonan et al., 2003). It was further determined that less than 1% became non-ambulatory due to injury in transit or upon arrival at their destination. Following inspection, 37% of the non-ambulatory dairy cattle were condemned (Doonan et al., 2003).

Definitions for deciding fitness or unfitness for travel of compromised cattle are vague, subjective, and therefore inconsistent. A questionnaire survey of Danish livestock haulers found that 35% were frequently in doubt about specific cows when it came to their assessments regarding fitness for travel. When asked about transport legislation regarding fitness for travel, only 52% of drivers were able to provide correct answers to the questions (Herskin et al., 2017).

An animal's 'fitness for transport' generally refers to their ability to withstand the stresses of transportation such that it will not cause further compromise of their welfare. This would include animals that are non-ambulatory, suffering severe lameness, in poor body condition and/or otherwise in poor health that would suggest they are unlikely to be able to withstand long-distance travel to a market or slaughter plant.

Lactating dairy cattle are likely to become very uncomfortable under circumstances where they cannot be milked at least 2 times per day. They should not be transported long distances or sold through markets where it may be several days before they reach a slaughter plant. Ambulatory animals with elevated body temperatures, drug residue associated with recent treatment, obvious fractures, severe injury or open wounds, blindness, cancer (lymphoma), unreduced prolapses (uterine, rectal or other) or animals in labor with the potential to calve during transport do not fit the criteria of 'ready for transport' and should not be transported (AABP Guidelines for Transportation and Fitness-to-Travel Recommendations for Cattle https://www.aabp.org/Reso urces/AABP_Guidelines/transportationguidelines-2019.pdf).

The Ontario Farm Animal Council's 'Caring for Compromised Cattle' document (http://gpvec.unl.edu/Elective_files/feedlot/Caring_Compromised _Cattle_ISU.pdf) suggests that decisions regarding an animal's suitability for transport be determined by producers asking themselves three basic questions prior to loading animals for transport: (1) Can the animal walk? (2) Will the animal be able to walk off the truck at the final destination? (3) If the animal were to be moved to slaughter, would they themselves feel safe consuming meat from the animal? A 'no' response to any of these questions means that operators would need to consider further treatment or on-farm euthanasia. Treatment is a reasonable option when it is likely to improve the animal's condition. If treatment is unlikely to result in recovery and human consumption would pose risks to human health, euthanasia is the better choice. Euthanasia procedures should conform to the American Veterinary Medical Association's 2020 guidelines for euthanasia (AVMA Euthanasia Guidelines 2020, https://ww w.avma.org/sites/default/files/2020-01/2020_Euthanasia_Final_1-15-20.pdf). The overriding consideration in these decisions is always the welfare of the animal.

7.3 Pre-planning to reduce cattle hauling risks

The process of assembling, loading, hauling and unloading cattle provides plenty of opportunity for additional distress and injury. Co-mingling with other animals, handling by people unfamiliar to them and placement in new or strange environments all compound the challenges animals face during transport. Cattle with painful conditions such as lameness whether it be a foot problem, or proximal limb injury are likely to endure even greater pain when transported. Postural responses to side-to-side, back to front or front to back movement causes fatigue and pain on painful joints and muscles (Cockram, 2019).

Anecdotal estimates are that 95% of cattle are transported by truck in the United States. Beef and dairy cattle operations often haul cull cows and bulls to nearby markets or slaughter plants in pick-up pulled gooseneck trailers. On occasion, large dairies and cattle operations will remove a sufficient number of animals from their herds to fill a semi-tractor trailer. In the United States, most beef calves travel from cow/calf operations to livestock markets, then on to feeding areas several states away. In some cases, large cow/calf operations are able to sell calves direct to feeding operations, thus bypassing markets. Dairy calves and heifers are sometimes raised at off-site locations several states away from the home farm.

It is very inefficient and costly for livestock haulers to make short or certainly long hauls without being fully loaded. This may require that they plan additional stops to ranches, feedlots or markets when and where cattle are likely to be assembled and ready to be loaded. Haulers need to know the approximate size and numbers of animals to be loaded to assure they will have adequate space and be able to distribute cattle on the trailer properly. If a hauler is moving fat cattle to a slaughter facility, they may be required to load cattle at a specific time of day or night while simultaneously arranging their travel schedule to arrive at a packing plant at a specific time of day in order to fit the queue or slaughter sequence at the abattoir. As slaughterhouses become scarcer, and the demand for cost-effective hauling services increases, the need for trip planning to maximize efficiency becomes greater and more complicated. Poor logistical management that leads to increased transport time affects morbidity, mortality, weight loss, carcass bruising and quality. Minimizing transport time and travel distance are primary objectives for preserving animal welfare.

7.4 Pre-planning for transport during inclement weather

It is not always possible to predict the potential for encountering weather extremes particularly when traveling significant distances. However, pre-planning based upon the best available weather information is a key step in the pre-transport process. If there is high likelihood of poor weather (heavy snow, for example), haulers may need to consider rerouting or delaying the trip until conditions improve. Drivers should anticipate that it would take longer to get to the final destination and determine what that might mean for the animals in transit.

Transporting animals during periods of extreme cold weather is not recommended. If cattle must be transported in cold and windy conditions, drivers should avoid stopping in order to arrive at their destination as quickly as possible. At temperatures of −0°F (−18°C), hauling cattle at highway speeds risks dangerous wind chill. The likelihood of animal death increased sharply when the midpoint ambient temperature fell below −15°C (5°F) (Gonzalez et al., 2012). The problem is compounded when cattle are wet. Steps to mitigate the cold temperature effects on animals would include the following measures. Cover bottom ventilation slats in the vehicle to protect cattle from cold air drafts and road spray, but allow for adequate ventilation at all times. Close nose vents in the trailer, supply ample bedding such as straw and avoid overcrowding to allow space for animals to reposition themselves to avoid the direct effects of wind and cold that might cause frostbite. Wet bedding should be removed and replaced with clean and dry bedding materials after each trip (Master Cattle Transporter Guide, https://www.bqa.org/media/bqa/docs/master_cattle_t ransporter_guide-digital.pdf).

Extreme heat and humidity can be deadly as well. The likelihood of animals becoming non-ambulatory increased when temperatures rose above 30°C (86°F) (Gonzalez et al., 2012). Temperature-humidity indexes equal to or greater than 100°F (38°C) pose extremely serious if not deadly risks to cattle (Knowles, 1999). If possible, avoid transporting animals on hot and humid days. However, if necessary, transport of cattle should be avoided during daylight hours between 11 AM and 4 PM. Drivers should avoid stopping, but when required they should look for shaded areas and keep stops as short as possible. As with cold weather conditions, reduce cattle density by loading fewer animals on the trailer. Extreme care in handling is especially critical under hot and humid weather conditions (Master Cattle Transporter Guide, https://www.bqa .org/media/bqa/docs/master_cattle_transporter_guide-digital.pdf).

7.5 Assembling, handling and loading

The separation and re-corralling of animals for the purpose of confinement in a pen or lot will naturally cause some degree of angst and discomfort. It is compounded when animals are confined or regrouped with new animals from other sources. In lactating cows, regrouping reduces milk production and time spent feeding and resting. Even minor changes cause variable degrees of aggression normally exhibited as dominance behavior. For example, fighting, in the form of head-to-head pushing will often continue for a period of 2–3 days while a new hierarchy is established (Phillips, 2002). While difficult to prevent these types of interactions, they can result in injury and carcass bruising. Normal patterns of feed and water consumption are also disrupted which is problematic for animals that may be transported long distances where access may be denied for 24 h or more.

Loading areas should have smooth solid sides free of protrusions that might cause injury. Floors should have secure non-slip footing which may include the use of rubber mats, sand and textured concrete surfaces. Handle cattle quietly during the loading process in order to avoid bruising or injury from slipping and falling. Chutes, alleyways, ramps and races should have secure footing, and solid sides free of protruding objects that could cause bruising. Loading ramps should be at an angle of less than 25° with non-slip flooring and no gaps between the ramp and the floor of the trailer.

Cull dairy cows that are lactating should be milked just prior to loading and transport. Be sure cattle have access to feed and water prior to transport to a market or distant packing plant and assure appropriate density on the trailer. Consider weather conditions and adjust animal density as necessary. Do not load cattle with conditions that are unlikely to pass ante-mortem inspection at slaughter including cancer eye and blindness, fever above 103°F (39°C), unreduced prolapses, severe emaciation and weakness, or cows in labor or where calving is imminent. Cows with lameness should be carefully scrutinized. Do not load cows with a locomotion score of 4 or 5 on a 5-point scale; rather consider the options of additional treatment or euthanasia.

8 Welfare of cattle during transport

8.1 Transport time and distance

The effect of the time a journey requires appears to be of greater significance than distance covered (Knowles, 1999; Gonzalez et al., 2012). Canadian researchers assessed the effect of long haul transport (> 400 km, i.e. 250 miles) on the welfare of cattle (Gonzalez et al., 2012). The incidence of dead, non-ambulatory and lame cattle, characterized as 'total compromised' animals, was determined using survey responses (n= 6,152) from drivers and hauling companies (16), involving 327 tractor-trailer units and 290 866 animals. The total number of compromised animals was 130 (0.045%–130/290 866); 32 (0.011%) died, 65 (0.022%) became non-ambulatory and 34 (0.012%) became lame in transit.

Other findings from this study were that cull cattle were at greatest risk of becoming lame at the time of loading and unloading, becoming non-ambulatory or dying during the journey compared with feeder calves and fat cattle. Observations also confirmed that the likelihood of death, becoming lame or non-ambulatory increased the longer animals were in transit. Death rate was higher for animals that lost more than 8% of body weight. The high loss of body weight was attributed to ambient air temperatures above 30°C (86°F) and a long period of time (> 30 h) on the truck (Gonzalez et al., 2012).

8.2 Stocking density

Despite industry proponents who argue for high stocking densities, research indicates that high stocking densities (Tarrant et al., 1988) reduce the welfare of animals during transport. When stocking densities are low and movement is less restricted, cattle are able to change positions frequently during the trip. The preferred orientation of cattle in low-density conditions is parallel to the direction of travel and loss of balance occurs primarily on gear changes, braking and cornering. In high stocking density conditions, movement is more restricted, thereby interfering with the animal's ability to adopt a preferred orientation to the direction of travel. Advocates of high stocking densities suggest that the opportunity to lean on an adjacent animal counteracts minor losses of balance, creating less potential for falling during transport.

The primary hazard in transport is for cattle that loose footing and go down during transport. Involuntary losses in balance that led to falling occurred more frequently in high stocking density conditions (Tarrant et al., 1988). Moreover, animals that went down in the trailer were subject to trampling and destabilization of the entire group that on occasion led to what authors' described as a 'domino effect'. Fear of falling is believed to be the primary reason cattle stand during transport and are likely to lie down only on very long journeys (Knowles, 1999). Overall, animal welfare and carcass quality are more adversely affected at higher stocking densities compared with medium and low stocking densities (Tarrant et al., 1988).

8.3 Behavior of cattle during transport

Cattle are naturally anxious and restless at the beginning of their journeys, which leads to frequent bouts of urination and defecation. Manure slurry and moisture reduces traction on trailer flooring surfaces and can increase the potential for falling. Social interactions characterized as head butting, mounting and pushing behavior are more commonly observed soon after loading, during the early part of the journey and when stocking densities are higher (Tarrant et al., 1988; Knowles, 1999). Exploratory behavior described as sniffing the environment, sexual and aggressive behaviors were reduced in high stocking density conditions.

8.4 Drivers and driving experience

Anyone who might think the job of a trucker who hauls a load of lumber is equivalent to hauling a load of livestock, should think again. Transporting live animals safely and on time to their destination can be extremely challenging. Livestock haulers must understand how to handle cattle and load a truck and trailer to distribute the weight appropriately. They must abide by schedules for cattle pickups at feedlots, farms as well as markets, and time these so that they can meet delivery schedules whether it be to another farm, market or to slaughter. Feedlot operators pay for losses that occur because of 'shrink'; therefore, producers will select haulers that can get cattle to packing plants with a minimum of weight loss. The health and welfare of cattle during transport is largely determined by the ability and experience of the driver.

Mortality rates for cattle during road transport are low compared to other species (Phillips, 2002; Gonzalez et al., 2012). More important are bruises that occur as a result of injuries or bruising associated with vehicle movement. Driving and driver experience are the most important factors with respect to injuries and bruising during transport (Knowles, 1999). Rapid starts, quick acceleration, braking, gear changes and excessive speeds while cornering are all causes for instability that increase the potential for bruising and in worst-case scenarios falling. Selecting routes that minimize the number of hills and winding roads can reduce these issues significantly.

In a long haul study by Gonzalez et al., the driver's experience had a significant impact on the welfare of cattle. Drivers with greater than 10 years of experience had a lower incidence of non-ambulatory and compromised cattle on arrival at their destinations compared to drivers of less than 10 years of experience. It is likely that experienced drivers not only have better driving skills, but also do a better job of distributing animals in the trailer. Researchers suggested that it may be advantageous to use experienced drivers when encountering weather extremes and journeys lasting more than 30 h (Gonzalez et al., 2012).

8.5 Livestock rollover accidents

There are few published reports regarding highway accidents involving livestock haulers. One that is frequently sighted is that by Woods and Grandin, 2008 (Woods and Grandin, 2008) of 415 commercial livestock truck accidents occurring in the United States and Canada between 1994 and June 2007. Fifty-nine percent of the accidents from this report occurred during the early morning hours from midnight to 9 AM. The majority, 80% of the accidents, involved a single vehicle. Driver error was cited as the cause for 85% of the accidents. In most cases (84%) the vehicle rolled over to the right side. Since vehicles travel on the right side of the road in North America, it was deduced that most cases result from the driver's fatigue and falling asleep while driving. Accidents were most common in the fall of the year and 69% of the trucks were hauling calves and yearlings to grower or feeder operations in excess of 500 km (310 miles). Most trips are made by a single driver with travel times of 20 h or more (Woods and Grandin, 2008).

8.6 The 28-hour law

The goal should always be to find markets and slaughter plants as close to the farm as possible. Journey time should be short enough to avoid the requirement to unload the animals for resting, which in the United States is 28 h. The 28-h law provides that animals cannot be transported by 'rail carrier, express carrier or common carrier' (except by air or water) for more than 28 consecutive hours without being unloaded for 5 h for rest, water and food (United States Code Annotated. Title 49. Transportation. Subtitle X. Miscellaneous. Chapter 805. Miscellaneous. § 80502. Transportation of animals. https://www.govinfo. gov/content/pkg/USCODE-2011-title49/pdf/USCODE-2011-title49-subti tleX-chap805-sec80502.pdf). If transport is anticipated to be equal to or more than 28 h, haulers must plan their routes so that rest periods may be accommodated. This rule, as good as it seems to be in principle, has been difficult if not impossible to enforce (Edwards-Callaway et al., 2019). Although it would seem logical to expect that a stop along the way in long-distance travel would be of value, there is also very little if any evidence to support benefits or reasonable guidelines.

8.7 Hours-of-service driving law

This law (developed during the Obama Administration in 2015) requires that truck drivers not be officially on duty for more than 14 h and while on duty, not have driven more than 11 of those hours. When drivers reach those maximum allotments of time, they are required to stop and rest for a minimum of 10 h. To assure compliance with the law, in December 2017 the Federal Motor Carrier Safety Administration (FMCSA) began requiring most US truckers to carry electronic logging devices (ELD) to verify that truckers were stopping to rest as required. The FMCSA, yielding to pressures from the agricultural commodities granted drivers an exemption from the requirement to use ELDs under certain travel conditions. As the law is currently implemented, cattle haulers are subject to the hours of service regulations when operating beyond 150 air miles from the source of the commodity; that is 11 h of driving, 14 h of on duty not driving and 10 consecutive hours off before proceeding to drive for 11 h. While operating within said 150 air miles, the hours of service rules do not apply. Drivers transporting livestock or insects are not subject to the ELD rule (https:/ /cms8.fmcsa.dot.gov/regulations/hours-of-service).

8.8 The 'Transporting livestock across America safely Act'

On January 10, 2019, Representative Ted Yoho (also a veterinarian) from Florida, introduced bill H.R. 487. 'Transporting Livestock Across America Safely Act' in the House of Representatives to develop a ruling that would be more acceptable to livestock haulers across the United States. Soon after its introduction to the House of Representatives, this bill was referred to the Subcommittee on Highways and Transit in February 2019 where it awaits further action. Among other provisions, this bill increases the drive time to a maximum of not less than 15 h and not more than 18 h within a 24-hour period. It also excludes time spent loading and unloading to more accurately include only travel time, which should make the rule more reasonable for livestock haulers (https://www.congress.gov/bill/115th-congress/senate-bill/2938/text?format =txt). No doubt, if this bill were likely to become law, it will undergo significant modification; nonetheless, efforts to move animals to their destinations quickly and safely are important for human health and animal welfare.

The 28-h law, 'Hours of Service Law' and 'Transporting Livestock Across America Safely Act' are intended to reduce the number of accidents involving over-the-road haulerswhile not causing unreasonable restrictions for livestock haulers. Nationwide, accident statistics indicate that truck-related fatalities have been on the rise in recent years. For example, in 2017, there were 841 occupants of large trucks killed in crashes, which was up from 725 in 2016, and 665 in 2015, according to a National Highway Traffic Safety Administration report Traffic Safety Facts 2017 Data – Large Trucks https://crashstats.nhtsa.dot .gov/Api/Public/ViewPublication/812663. The most common causes of truck accidents are fatigue, impaired and distracted drivers. Meaningful laws that can promote highway safety are good for humans and animals alike.

9 Welfare of cattle at slaughter

The welfare of cattle at slaughter is determined by the evaluation of animal-based measures assessed on arrival, while in lairage, during movement to the stunner and at the point of stunning, exsanguination and death. Specific outcome measures include an enumeration of the following: animals dead or non-ambulatory on arrival; the number of animals observed to slip or fall during unloading, in lairage, or during movement to the stunner; evidence of vocalization, amount of prod use, time from stunning to exsanguination ('stun to stick' time) and effectiveness of stunning procedures.

9.1 Unloading

The first step on arrival at a packing plant is unloading. Under ideal conditions, this should occur within 15 min and hopefully no more 60 min after arrival.

Trucks are normally scheduled so that delays are minimized which is critically important in hot weather to avoid heat stress. Cattle should exit the truck or trailer calmly with little or no encouragement. Industry guidelines recommend that less than 5% of animals be unloaded with the use of electric prods and less than 1% of animals fall during the unloading process (Grandin, 2005a, https ://www.grandin.com/RecAnimalHandlingGuidelines.html). Unloading areas should have secure flooring surfaces and provide a sufficiently wide and clear path to the holding pen or lairage area. Concrete surfaces may be grooved in a series of parallel grooves or in a diamond pattern to increase traction on concrete floors. Cattle should be carefully observed at unloading for evidence of reduced mobility or other condition (2016 Guidelines for Humane Slaughter, https://www.avma.org/sites/default/files/resources/Humane-Slaughter-Guideli nes.pdf).

9.2 Mortalities and non-ambulatory cattle on arrival at slaughter

The number and percentage of animals dead or non-ambulatory on arrival at slaughter plants should be monitored. Both conditions are most likely the result of the shipment of animals that were not fit for travel. In the United States, non-ambulatory disabled cattle are defined as any animal that cannot rise from a recumbent position or cannot walk. Such cattle cannot enter a packing plant. This includes animals with broken legs, severed tendons or ligaments, paralysis from nerve injury, and fractures of the vertebral column or metabolic conditions. This also applies to cattle demonstrating neurological symptoms even if they are ambulatory.

At one time, the Food Safety and Inspection Service (FSIS) would permit the slaughter of cattle under 'emergency conditions'. For example, an animal that had passed ante-mortem inspection, but later sustained an acute injury in route to the knock box (cattle stunner) could still be slaughtered. The agency reversed its position on emergency slaughter in early 2008 following the inhumane handling of non-ambulatory animals at a California slaughter plant. That incident resulted in the recall of 143 million pounds of fresh and frozen beef products dating back to February 2006 (the largest meat recall in US history). Regulations now prohibit the slaughter of any non-ambulatory animal arriving at a slaughter plant (Becker, 2009 https://nationalaglawcenter.org/wp-content/uploads/assets/crs/RS22819.pdf .

9.3 Severe lameness and emaciation

Regulations on slaughter of cattle also extend to animals with severe lameness and emaciation. Nicholson et al. (2013), observed that 2.7% of cull dairy cows were severely lame (i.e. displaying an arched back at all times and refusing to bear weight on one leg), 4% of cattle were extremely emaciated and 4% had foot abnormalities (Nicholson et al, 2013). If an animal is too weak or lame to freely walk off the truck, it will likely have difficulty walking to the stunner and is therefore unlikely to pass ante-mortem inspection. A survey of 10 cattle auction markets in the western United States, found that 15.1% beef cows, and 15.4% of beef bulls were lame. Observations were worse on dairy cows, with lameness found in 44.7% of dairy cows and 26.1% of dairy bulls. This survey also found severe emaciation (i.e. dairy BCS = 1 and beef BCS = 1) in 13.3% of cull dairy and 3.9% of cull beef cows (Ahola et al., 2011).

9.4 Lairage

Cattle should be maintained with the animals in which they arrived. This is necessary to reduce the potential for fighting and dominance behavior that increases stress in cattle and the amount of carcass bruising and dark cutting beef. Bulls are prone to fighting and mounting and may need to be separated to avoid excessive bruising and dark cutters. Lairage pens should be of sufficient size to accommodate a full truckload of cattle. Pens should have water troughs and sufficient space so that animals are able to lie down, rest and allow observation by veterinary inspectors during ante-mortem inspection (AVMA Humane Slaughter Guidelines, 2016).

9.5 Movement to the stunner

As cattle move from lairage pens to the lead up alley heading to the stunning area, there should be no slipping, falling or vocalization. Alleyways should be free of extraneous debris that might cause distraction and hesitation in the forward movement of cattle. Animals prefer to go from dark areas to areas that are well lit; therefore, lighting that eliminates shadows improves cattle flow and reduces balking. Animal handlers must understand the principles of flight zone and point of balance for low stress handling of cattle. Handlers should also limit the number of animals in crowd pen areas to no more than 3-5.

Plants with good handling practices should be able to have less than 1% of animals falling and less than 3% vocalizing from adverse handling. Use of electric prods should be minimized by restricting use and availability of the devices to trained personnel and only in areas where animals are likely to balk (i.e. prior to entering the stun box) (AVMA Guidelines for Humane Slaughter 2016).

9.6 Restraint at the stun box

In packing plants where a stun box is used, pressure used to restrain the animal should be sufficient for restraint but not to the extent, that it causes pain or distress. Cattle are prone to balk upon entry to the stunning box, which may require the need for a prod or push gate. Improved designs move the front of the stun box assembly, including the head holder, backward toward the animal. This eliminates the need to prod or push the animal forward. Large plants use a center V track restrainer system, which maintains animals in an upright comfortable position while moved to the stunner on a conveyor. These systems also facilitate stunning by positioning the stunner operator in a safer and more effective position for proper anatomic placement of the captive bolt (AVMA Guidelines for Humane Slaughter, 2016).

9.7 Captive bolt stunning

The physical methods used for the humane stunning of cattle are penetrating and non-penetrating captive bolt and firearms. The objective of these devices is to cause an immediate loss of consciousness by causing trauma to the cerebral hemispheres and brainstem. Standing animals collapse immediately with a properly placed shot, followed by muscular rigidity and eventually relaxation that leads to a period of uncoordinated limb movement. There is an immediate cessation of rhythmic breathing and loss of corneal and palpebral reflexes.

There are two types of captive bolt, penetrating and non-penetrating. Penetrating captive bolt discharges a bolt through the skull and into the brain. Non-penetrating captive bolt is equipped with a convex mushroom-shaped head that causes unconsciousness by concussion, fracture of the skull and focal damage to the brain around the site of impact. Captive bolts are powered by either compressed air or gunpowder cartridge charges ranging from 9 mm, 22 caliber or 25 caliber. Dysfunction of powder-charged captive bolt guns is primarily due to a lack of maintenance or damp cartridges from improper storage of powder charges (AVMA Guidelines for Humane Slaughter, 2016).

Properly applied by a well-trained operator, captive bolt can easily render 95% or more of animals unconscious with a single shot (Grandin, 2002). Shot effectiveness below 95% is indicative of a problem with either the person doing the stunning or the stunning equipment. Grandin reports that the best plants have a 99% first shot efficacy (Grandin, 2005b). Head holders can be used to improve shot accuracy, but they must be properly designed or they may cause additional stress on restrained animals. The head remains stable without restraint on center-track conveyor systems, which helps to increase the accuracy of shot placement (2016 AVMA Humane Slaughter).

9.8 Stun to stick interval

Captive bolt stunning is intended to cause an immediate and sustained loss of consciousness persisting until exsanguination reduces blood pressure and volume sufficient to cause death. Prolongation of the time from stunning to exsanguination (better known as the 'stun-to-stick interval') risks the possibility of an animal regaining consciousness. The recommended maximum stun-to-stick times for cattle are 60 s for penetrating and 30 s for non-penetrating captive bolt (Captive-Bolt Stunning of Livestock (2013) Humane Slaughter Association, United Kingdom: https://www.hsa.org.uk/downloads/publications/captivebo ltstunningdownload.pdf).

Exsanguination of sheep and goats is typically performed by making a ventral incision in the neck severing the carotid arteries and jugular veins. In cattle, exsanguination is performed by inserting a sharp knife into the jugular furrow at the base of the neck. Once the knife is through the skin it is directed toward the thoracic inlet where severing of the large arteries and veins at the base of the neck assures rapid loss of blood and death (Captive-Bolt Stunning of Livestock (2013) Humane Slaughter Association, United Kingdom: https://www.hsa.org.uk/downloads/publications/captiveboltstu nningdownload.pdf).

9.9 Firearms

Gunshot is the most practical for slaughter of animals with heavy skulls such as large bulls, boars and bison. Safety is the major concern when firearms are used for slaughter or euthanasia purposes. Selection of a firearm and bullet that is sufficient to traverse the skull, but not perforate (i.e., pass through) the skull is desired. A properly placed bullet will cause massive brain damage and death with little likelihood of a return to consciousness.

Handguns in calibers ranging from .32 to .45 caliber, rifles from .22 magnum or higher for thick skulled bulls, boars and bison and shotguns gauges .410, 28, 20, 16 or 12 are acceptable. When using a handgun, rifle or shotgun at close range, hold the muzzle of the firearm within 3 feet of the intended target. Solid point bullets or hollow points with controlled expansion properties are best for handguns and rifles. Number 6 or larger birdshot is acceptable for shotguns; buckshot or slugs are acceptable but probably excessive since most shots are within close range (Shearer, 2018). Full metal-jacket bullets are not recommended for use in slaughter plants for safety reasons (AVMA Guidelines for Humane Slaughter, 2016).

An important difference with the use of a firearm as compared with captive bolt is that the muzzle of a firearm should never be held flush with the skull. Enormous pressures are developed within the barrel when it is discharged that could cause explosion of the barrel putting the shooter and by-standers at great risk. Captive bolt on the other hand, requires immediate contact with the skull otherwise, one is unlikely to realize the full benefit of bolt extension into the skull.

9.10 Anatomic landmarks for the use of the penetrating captive bolt and gunshot

Selection of the ideal anatomic site for use of a captive bolt or firearm is a key prerequisite to stunning success. In cattle, the point of entry of the projectile should be at the intersection of two imaginary lines, each drawn from the outside corner of the eye to the center of the base of the opposite horn (2020 AVMA Euthanasia Guidelines; AVMA Guidelines for Humane Slaughter, 2016). Alternatively, in long-faced cattle or young-stock, a point on the midline of the forehead that is halfway between the top of the poll and an imaginary line connecting the outside corners of the eyes can be used (2016 AVMA Humane Slaughter Guidelines). Firearms should be positioned so that the muzzle is perpendicular to the skull to direct the projectile toward the brainstem (Fig. 3).

Proper anatomic site selection and use of the appropriate caliber handgun, rifle or gauge of shotgun and bullets/shells assures effective results. A well-trained shooter should be able to effectively assure unconsciousness and death

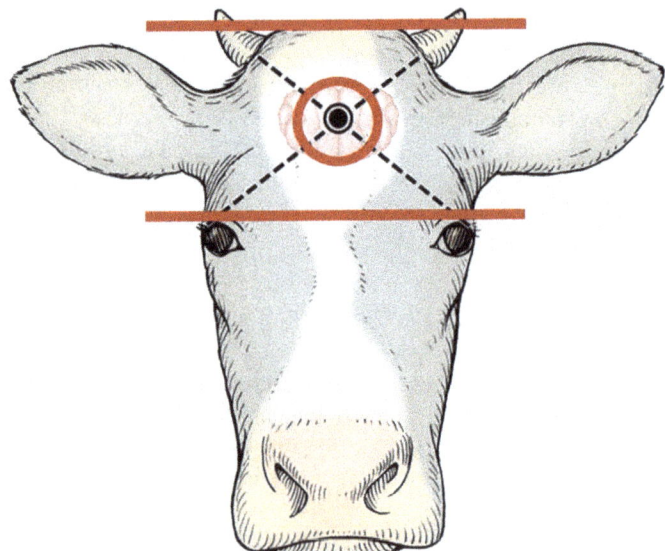

Figure 3 Guidelines for anatomic site placement of a captive bolt or bullet from a firearm. The point of entry of the projectile should be at the intersection of two imaginary lines (represented as the black dashed lines), each drawn from the outside corner of the eye to the center of the base of the opposite horn. This site can also be determined as a point on the midline of the face (defined by the red circle) that is halfway between the top of the poll and an imaginary line connecting the outside corners of the eyes can be used (adapted from the website of Shearer and Ramirez, Humane Euthanasia of Livestock, Iowa State University; AVMA Guidelines for the Humane Slaughter of Animals: 2016 Edition; AVMA Guidelines for the Euthanasia of Animals: 2020 Edition).

in 95% or more of the animals with a single shot (AVMA Guidelines for the Humane Slaughter of Animals: 2016 Edition https://www.avma.org/sites/def ault/files/resources/Humane-Slaughter-Guidelines.pdf accessed on January 2020.)

10 Summary

Transport is an extremely stressful event for cattle, but there are ways to reduce its impacts on their welfare. Dairy cattle typically leave herds at less than 5 years of age, whereas beef cattle are roughly 10 or more years of age. Most dairy and beef herds sell cattle through livestock markets. Compromised cattle should be transported directly to the nearest slaughter facilities. Dairy cattle arriving at slaughter facilities are frequently light muscled and in poor body condition. This increases the potential for bruising that causes greater economic loss from the need to trim away bruised tissues. The incidence of visually obvious physical defects in cattle is low; however, more than one-third of hides are damaged by branding. Carcass evaluation at slaughter plants demonstrates a high incidence of carcass bruising and dark cutting beef, in cases where animals have experienced a significant degree of pre-slaughter stress. These observations confirm the fact that transport and handling at slaughter are major welfare issues. Improving the welfare of animals and quality of beef from animals during transport and at slaughter begins at the pre-transport phase. Determining fitness for travel and pre-planning that takes into consideration the condition of animals to be transported, time and distance to the desired destination and weather conditions are key steps in optimizing welfare of animals that must be transported. The natural behavior of cattle is to stand while in transit. In long-distance travel, this increases fatigue and potential for injury, particularly for animals compromised by lameness, emaciation or other condition. Cattle should be carefully observed upon arrival at slaughter. If an animal is non-ambulatory it must be humanely euthanized. Animals that exit the trailer freely should be moved to lairage that offers immediate access to water and sufficient space for animals weary from travel an opportunity to lie down and rest. The plant's slaughter schedule should be managed such that cattle spend minimal time in lairage prior to movement to stunning and slaughter. Animals should be moved calmly and with a minimum of prodding in route to stunning and slaughter. Upon entry into the stun box, stunning should proceed without interruption. Personnel delegated the task of stunning should be able to effectively render animals unconscious with 95% or better accuracy with the first shot. If a second shot is required, it should be applied immediately. Facilities and movement of animals from the stun box to the bleeding rail should be designed to assure a maximum of 60 s from stunning to the point of exsanguination. Use of the appropriate anatomic site assures optimal success and animal welfare at slaughter.

11 Where to look for further information

There is an abundance of information available on the subject of livestock transport. Several key references for those desiring additional information are included in the list of references that accompany this article. However, consolidation of the livestock industry and the reduction in availability of livestock market and packing plants throughout North America will continue to create challenges for ensuring the welfare of animals that require transport. Interest in the ability to improve pre-transport preparation, to reduce stress during transport and to better manage conditions occurring in the post-transport phase will continue to be priorities for future research.

There are multiple organizations with a stake in assuring the welfare of cattle during transport including the National Cattlemen's Beef Association (NCBA), Livestock Market Association (LMA), National Milk Producers Federation (NMPF) and the North American Meat Institute (NAMI) to mention just a few. However, beyond producer, market and meat packers are consumers. Now, more than at any time in history, consumers want to know that the welfare of animals is assured on the farm, at markets at slaughter and points in between.

12 References

Ahola, J. K., Foster, H. A., VanOverbeke, D. L., Jensen, K. S., Wilson, R. L., Glaze, J. B., Fife, T. E., Gray, C. W., Nash, S. A., Panting, R. R. and Rimbey, N. R. 2011. Survey of quality defects in market beef and dairy cows and bulls sold through livestock auction markets in the Western United States: I. Incidence rates. *J. Anim. Sci.* 89(5):1474–1483.

American Association of Bovine Practitioners. 2019. AABP Guidelines for Transportation and Fitness-To-Travel Recommendations for Cattle. Available at: https://www.aabp .org/Resources/AABP_Guidelines/transportationguidelines-2019.pdf.   Accessed January 2020.

American Veterinary Medical Association. 2016. AVMA Guidelines for the Humane Slaughter of Animals: 2016 Edition. Available at: https://www.avma.org/sites/def ault/files/resources/Humane-Slaughter-Guidelines.pdf. Accessed January 2020.

American Veterinary Medical Association. 2020. AVMA Guidelines for the Euthanasia of Animals: 2020 Edition. Available at: https://www.avma.org/sites/default/files/2020 -01/2020_Euthanasia_Final_1-15-20.pdf. Accessed January 2020.

Becker, G. S. 2009. *Nonambulatory Livestock and the Humane Methods of Slaughter Act.* Congresional Research Service. 7-5700, www.crs.gov, RS22819. p. 1–7.

Bethancourt-Garcia, J. A., Vaz, R. Z., Vaz, F. N., Silva, W. B., Pascoal, L. L., Mendonça, F. S., Vara, C. Cd, Nuñez, A. J. C. and Restle, J. 2019. Pre-slaughter factors affecting the incidence of severe bruising in cattle carcasses. *Livest. Sci.* 222:41–48.

Cockram, M. S. 2019. Fitness of animals for transport to slaughter: a Review. *Can. Vet. J.* 60(4):423–429.

De Vries, A. 2013. Cow longevity economics: the cost benefit of keeping the cow in the herd. Proceedings of the From the Cow Longevity Conference 2013, Hamra Farm, Sweden. Available at: http://www.milkproduction.com/Library/Scientific-articles/Management/Cow-longevity-economics-The-cost-benefit-of-keeping-the-cow-in-the-herd/. Accessed December 2019.

Doonan, G., Appelt, M. and Corbin, A. 2003. Nonambulatory livestock transport: the need for consensus. *Can. Vet. J.* 44(8):667–672.

Edwards-Callaway, L. N., Walker, J. and Tucker, C. B. 2019. Culling decisions and dairy cattle welfare during transport to slaughter in the United States. *Front. Vet. Sci.* 5: :1–5. Article 343.

Geiger, C. 2016. What is the national cull rate? Hoard's Dairymen Intel. Available at: https://hoards.com/article-19040-what-is-the-national-cull-rate.html. Accessed December 2019.

Gonzalez, L. A., Schwartzkopf-Genswein, K. S., Bryan, M., Silasi, R. and Brown, F. 2012. Relationships between transport conditions and welfare outcomes during commercial long haul transport of cattle in North America. *J. Anim. Sci.* 90(10):3640–3651.

Grandin, T. 2002. Return-to-sensibility problems after penetrating captive bolt stunning of cattle in commercial beef slaughter plants. *J. Am. Vet. Med. Assoc.* 221(9):1258–1261.

Grandin, T. 2005a. Edition, with 2007 and 2010 Updates. Recommended Animal Handling Guidelines and Audit Guide for Cattle, Pigs, and Sheep. Available at: https://www.grandin.com/RecAnimalHandlingGuidelines.html. Accessed January 2020.

Grandin, T. 2005b. Maintenance of good animal welfare standards in beef slaughter plants by use of auditing programs. *J. Am. Vet. Med. Assoc.* 226(3):370–373.

Grandin, T. n.d. How to Track Down the Cause of Bruising. Available at: https://www.grandin.com/references/cause.bruising.html. Accessed December 26, 2019.

Grandin, T. 2013. *Recommended Animal Handling Guidelines and Audit Guide: A Systematic Approach to Animal Welfare.* AMI Foundation, Washington, DC. Available at: https://www.grandin.com/RecAnimalHandlingGuidelines.html. Accessed 20 December 2019.

Harris, M. K., Eastwood, L. C., Boykin, C. A., Arnold, A. N., Gehring, K. B., Hale, D. S., Kerth, C. R., Griffin, D. B., Savell, J. W., Belk, K. E., Woerner, D. R., Hasty, J. D., Delmore, R. J., Martin, J. N., Lawrence, T. E., McEvers, T. J., VanOverbeke, D. L., Mafi, G. G., Pfeiffer, M. M., Schmidt, T. B., Maddock, R. J., Johnson, D. D., Carr, C. C., Scheffler, J. M., Pringle, T. D. and Stelzleni, A. M. 2017. National beef quality audit–2016: transportation, mobility, live cattle, and carcass assessments of targeted producer-related characteristics that affect value of market cows and bulls, their carcasses, and associated by-products. *Translat. Anim. Sci.* 1(4):570–584.

Herskin, M. S., Hels, A., Anneberg, I. and Thomsen, P. T. 2017. Livestock drivers' knowledge about dairy cow fitness for transport – a Danish questionnaire survey. *Res. Vet. Sci.* 113:62–66.

Hours of Service Rule. 2020. U.S. Department of Transportation. Federal Motor Carrier Safety Administration. Available at: https://cms8.fmcsa.dot.gov/regulations/hours-of-service. Accessed January 2020.

HSA. 2013. Captive-Bolt Stunning of Livestock. Humane Slaughter Association, United Kingdom. Available at: https://www.hsa.org.uk/downloads/publications/captiveboltstunningdownload.pdf. Accessed January 2020.

Huertas, S. M., Gil, A. D., Piaggio, J. M. and van Eerdengurg, F. J. C. M. 2010. Transportation of beef cattle to slaughterhouses and how this relates to animal welfare and carcase bruising in an extensive production system. *Anim. Welf.* 19:281–285.

Jarvis, A. M., Selkirk, L. and Cockram, M. S. 1995. The influence of source, sex class and pre-slaughter handling on the bruising of cattle at two slaughterhouses. *Livest. Prod. Sci.* 43(3):215–224.

Knowles, T. G. 1999. A review of the road transport of cattle. *Vet. Rec.* 144(8):197–201.

Lee, T. L., Reinhardt, C. D., Bartle, S. J., Vahl, C. I., Siemens, M. and Thomson, D. U. 2017. Assessment of risk factors contributing to carcass bruising in fed cattle at commercial slaughter facilities. *Transl. Anim. Sci.* 1(4):489–497.

Livestock Trailer Safety. 2012. Farm and Ranch eXtension in Safety and Health (FReSH) Community of Practice. Available at: http://www.extension.org/pages/64391/liv estock-trailer-safety. Accessed December 2019.

Master Cattle Transporter Guide, National Beef Quality Assurance Guide for Cattle Transporters. Available at: https://www.bqa.org/Media/BQA/Docs/master_cattle_t ransporter_guide-digital.pdf. Accessed December 2019.

Matulis, R. J., McKeith, F. K., Faulkner, D. B., Berger, L. L. and George, P. 1987. Growth and carcass characteristics of cull cows after different times-on-feed. *J. Anim. Sci.* 65(3):669–674.

National Cattlemen's Beef Association. 2017. *National Beef Quality Audit-2017: Market Cows and Bulls, Executive Summary*. National Cattlemen's Beef Association, Centennial, CO.

National Highway Traffic Safety Administration report Traffic Safety Facts 2017. Data – Large Trucks. National Center for Statistics and Analysis. (2019, January). Large trucks: 2017 data. Washington, DC: National Highway Traffic Safety Administration. Available at: https://crashstats.nhtsa.dot.gov/Api/Public/ViewPublication/812663. Accessed January 2020.

Nicholson, J. D. W. 2008. National Market Cow and Bull Beef Quality Audit-2007: a survey of producer-related defects. M.S. Thesis, Texas A&M University, College Station.

Nicholson, J. D. W., Nicholson, K. L., Frenzel, L. L., Maddock, R. J., Delmore, R. J., Lawrence, T. E., Henning, W. R., Pringle, T. D., Johnson, D. D., Paschal, J. C., Gill, R. J., Cleere, J. J., Carpenter, B. B., Machen, R. V., Banta, J. P., Hale, D. S., Griffin, D. B. and Savell, J. W. 2013. Survey of transportation procedures, management practices, and health assessment related to quality, quantity, and value for market beef and dairy cows and bulls. *J. Anim. Sci.* 91(10):5026–5036.

Ontario Farm Animal Council. 2010. The Ontario Farm Animal Council's "Caring for Compromised Cattle" document. Available at: http://gpvec.unl.edu/Elective_files/feedlot/Caring_Compromised_Cattle_ISU.pdf. Accessed January 2020.

Phillips, C. 2002. *Cattle Behaviour and Welfare* (2nd edn.), *Chapter 5: The Welfare of Cattle during Transport, Marketing and Slaughter*. Wiley-Blackwell, Hoboken, NJ. p. 38–48.

Ponnampalam, E. N., Hopkins, D. L., Bruce, H., Li, D., Baldi, G. and Bekhit, A. E. 2017 Causes and contributing factors to "Dark Cutting" meat: current trends and future directions: a review. *Compr. Rev. Food Sci. Food Saf.* 16(3):400–430.

Sánchez-Hidalgo, M., Rosenfeld, C. and Gallo, C. 2019. Associations between pre-slaughter and post-slaughter indicators of animal welfare in cull cows. *Animals* 9(642):1–15.

Schnell, T. D., Belk, K. E., Tatum, J. D., Miller, R. K. and Smith, G. C. 1997. Performance, carcass, and palatability traits for cull cows fed high-energy concentrate diets for 0, 14, 28, 42, or 56 days. *J. Anim. Sci.* 75(5):1195–1202.

Shearer, J. K. 2018. Euthanasia of cattle: practical considerations and application: a review. *Animals* 8(57):1–17.

Stalsberg, K., Halfman, B., Stuttgen, S. and Skjolaas, C. A. 2017. *Livestock Hauling BQA Style, Wisconsin Beef Quality Assurance*. University of Wisconsin-Extension. Available at: https://fyi.extension.wisc.edu/wbic/files/2018/03/3T-handout-1done.pdf. Accessed December 2019.

Tarrant, P. V., Kenny, F. J. and Harrington, D. 1988. The effect of stocking density during 4-hour transport to slaughter on behaviour, blood constituents and carcass bruising in Friesian steers. *Meat Sci.* 24(3):209–222.

The Congress of the United States. 2018. Transporting Livestock Across America Safely Act. Available at: https://www.congress.gov/bill/115th-congress/senate-bill/2938/text?format=txt. Accessed January 2020.

Tucker, C. B., Mintline, E. M., Banuelos, J., Walker, K. A., Hoar, B., Varga, A., Drake, D. and Weary, D. M. 2014. Pain sensitivity and healing of hot-iron cattle brands. *J. Anim. Sci.* 92(12):5674–5682.

United States Code Annotated. 1994. Title 49. Transportation. Subtitle X. Miscellaneous. Chapter 805. Miscellaneous. § 80502. Transportation of animals (28 hour rule). Available at: https://www.govinfo.gov/content/pkg/USCODE-2011-title49/pdf/USCODE-2011-title49-subtitleX-chap805-sec80502.pdf. Accessed December 2019.

USDA. 2010. Beef 2007–08 Part IV: Reference of Beef Cow-calf Management Practices in the United States, 2007–08. Available at: https://www.aphis.usda.gov/animal_health/nahms/beefcowcalf/downloads/beef0708/Beef0708_dr_PartIV_1.pdf. Accessed December 2019.

USDA. 2018. Dairy 2014, Health and Management Practices on U.S. Dairy Operations, 2014. Report number 3, USDA-APHIS-VS-CEAH-NAMHS, Ft. Collins, CO. Available at: https://www.aphis.usda.gov/animal_health/nahms/dairy/downloads/dairy14/Dairy14_dr_PartIII.pdf. Accessed December 2019.

Woods, J. and Grandin, T. 2008. Fatigue: a major cause of commercial livestock truck accidents. *Vet. Ital.* 44(1):259–262.

Chapter 2

Ensuring the welfare of culled dairy cows during transport and slaughter

Carmen Gallo and Ana Strappini, Animal Welfare Programme, Faculty of Veterinary Science, Universidad Austral de Chile, Chile

1 Introduction

2 Legislation and codes of practice

3 Pre-transport conditions that influence the welfare of cows during transport

4 Welfare of culled cows during transport

5 The effects of livestock markets on cow welfare

6 Welfare of cows at the slaughter plant

7 Conclusions

8 Where to look for further information

9 References

1 Introduction

Culling is the departure of cows from the herd because of sale, slaughter, salvage or death (Fetrow et al., 2005). The proportion of cows that are culled from dairy herds annually is variable (25–30%), and depends on biological factors such as age, parity, milk yield, reproductive and sanitary state of the cows, and also on economic factors such as the price of milk, of cows and replacement heifers (Bascom and Young, 1998). Most of the culled cows will be either sold through livestock markets and livestock dealers or sent directly to slaughterhouses. According to the studies of González et al. (2012a,b,c) in cattle submitted to long haul in North America, for all loads of different categories surveyed, culled cattle represented only 0.9%. In Chile, Gallo et al. (1999) registered the different categories of cattle arriving at several Chilean slaughterhouses, finding that adult and old cows represented around 15% of all cattle slaughtered; most of these cows were actually slaughtered at the

http://dx.doi.org/10.19103/AS.2016.0006.05

smallest, and not exporting, meat plants. Another survey in Chile described 413 transport loads arriving at slaughterhouses and showed that 9.4% of them were culled cows (Gallo et al., 2005). Clearly, the number of culled cows slaughtered yearly in each country is variable and will depend mostly on economic factors, but their proportion with regard to the total of all cattle categories slaughtered is rather low. However, the sale of market cows and bulls accounts for 25% of all U.S. beef consumption (U.S., 2007). Culled cows are part of the food supply chain and should be treated accordingly to provide meat that is acceptable from a meat safety and quality point of view, as well as from an ethical point of view that considers animal welfare. Good handling during transport and slaughter aims to achieve that.

2 Legislation and codes of practice

Risks during transport can be reduced by selecting animals best suited to cope with the ordeals of journey. Therefore, the World Organisation for Animal Health standards (OIE, 2015) state that each animal should be inspected in order to evaluate fitness for transport and those that are considered unfit for transport should not be loaded, unless they are sent for veterinary treatment. Among animals unfit for transport are those sick, injured, weak, disabled or fatigued; those that are unable to stand unaided or cannot be moved without causing them additional suffering; those whose physical condition would result in poor welfare; females travelling without their young ones, that is, females that have given birth within the previous 48 h and pregnant animals that would be in the final 10% of their gestation period at the planned time of unloading. In accordance with OIE standards, all revised legislation/codes of practice (USA, 2003; European Council, 2005; New Zealand, 2011; Australia, 2012; Chile, 2013; Canada, 2015) indicate that animals should be checked before transport on their fitness for the journey; if not fit they should not be loaded and transported, although in some cases they may be acceptable with special provisions. Regarding competence of drivers, only in New Zealand, EU and Chile, a certificate of competence is required (Table 1).

All animals should be transported for the shortest possible time (OIE, 2015). Hartung et al. (2003) propose that the welfare of the animals is limited by their needs and not by a fixed maximum transport time if vehicle and transport conditions are appropriate. Accordingly, the OIE standard for the transport of animals by land (OIE, 2015) recommends that suitable water and feed be made available as appropriate and needed for the species, age and condition of the animals, as well as the duration of the journey, climatic conditions, etc. and hence does not refer to maximum transport duration. Within the revised legislation, Table 1 shows that there are maximum continuous journey times stated for cattle, in most cases separated according to two cattle classes:

Table 1 Maximum journey duration, space availability and other indications for cattle transport in different countries

Indications	Australia	Canada	New Zealand	EU	Chile	USA
Fitness for transport	✓	✓	✓	✓	✓	✓
Drivers' competence	NS	NS	✓	✓	✓	NS
Max journey duration (h)	24[a] 48[b]	18[a] 48[b]	12[a] 24[b]	8 (14-1-14)	NS 24	NS 28
Max time without water/food	24[a], 48[b]/NS	52/52	12[a] 24[b] /24[a], 48[b]	14/14	24/24	NS

Space availability:

Australia

Mean live weight (kg)	Minimum floor area (m²/head) standing
300	0.86
350	0.98
400	1.05
450	1.13
500	1.23
550	1.34
600	1.47
650	1.63

Canada

Weight range (kg)	Minimum space (m²/head)*
315-360	0.948
360-405	1.050
405-450	1.133

*Recommendations for safe load levels for the transportation of cattle by road from Canadian code of practice for beef cattle

New Zealand

Mean live weight (kg)	Minimum floor area (m²/head)
300	0.86
400	1.06
500	1.27
>600	1.50

EU

Mean liveweight (kg)	Minimum floor area (m²/head) standing
325	0.95-1.30
550	1.30-1.60
>700	1.47
650	>1.60

Chile: 1m²/500 kg

USA: NS

a = Young (1-6 months), lactating or pregnant cattle, b = others, NS = not specified.

young (one to six months), pregnant or lactating; and others (adult cattle). Also, maximum times without water/food are specified in most cases. Usually, the maximum journey time is equivalent to the maximum time without water and food; therefore, there are spells for unloading, resting, watering and feeding the animals, unless the vehicle is equipped with systems for watering and feeding. In these cases, space availability per animal must be increased.

The OIE (2015) standard for the transport of animals by land also indicates that the space required on a vehicle or in a container depends upon whether or not the animals need to lie down or stand; when they lie down, they should be able to adopt a normal lying posture and when they are standing they should have enough space to adopt a balanced position. Again, space recommendations for transport should be stated according to the needs of the animals. One criterion for acceptable stocking densities is based on the provision of adequate ventilation and another is the minimum space required for animals based on dimensions and activities during transport (Randall, 1993). Existing legislation for the transport of cattle in most countries provides general guidelines on space requirements for cattle and refers separately to different weights and sometimes classes of cattle (Table 1). This gives minimum space availabilities that range from 0.86 m^2/head for 300 kg cattle to 1.6 m^2/head for 650 kg cattle, which is based on Randall (1993) equation $A = 0.01 W^{0.78}$ (A = area, W = weight of animal) for standing cattle. For long distance transport, the FAWC (2013) recommends somewhat higher space allowances by using the equation $A = 0.021 W^{0.67}$. In the case of Chilean legislation, only one figure for minimum space availability is given (1.0 m^2/500 kg cattle), and hence, there is no difference due to cattle category; as it is common in many countries, there is a tendency towards using these minimum space availabilities commercially (Gallo et al., 2005; Gallo and Tadich, 2008).

In conclusion, there are recommendations in various countries in terms of minimum space allowances for different categories of cattle and also maximum transport times without food and water, but most given values are indistinctive of animal factors like breed, age, sex, body condition score, physiological state (pregnant/non-pregnant, lactating, weaned/not weaned, etc.), physical conditions (horned/not horned) or according to ambiental factors like expected length of journey, expected climatic conditions/temperature, ventilation during the journey and whether food and water need to be supplied during transport. All these factors should be taken into consideration when transporting cattle, and these become more important in animals that, although considered fit for transport, are often in poor physical condition, pregnant, lactating or even suffering from a painful illness (lameness, mastitis). As this is often the case with culled cows, special provisions should be made for their transport in order to reduce risks of poor animal welfare.

3 Pre-transport conditions that influence the welfare of cows during transport

Compared to the dairy industry, the physical condition of culled beef cows is better than that of dairy cows sent for slaughter (Grandin, 2001). When dairy cows are culled due to reproductive problems or even because they are difficult to handle, they may be in good physical shape and this may not represent a major welfare problem. However, often dairy cows are culled due to lameness, mastitis, milk fever or metabolic disorders that cause a poor physical condition, and hence, they may be too weak for transport. According to Grandin (1998, 2001), this happens when cows are pushed beyond their biological limits, and therefore, emphasis must be on preventing cows from becoming non-ambulatory.

All of these conditions will frequently apply to culled dairy cows. Moreover, Riehn et al. (2014) evaluated the proportion of pregnant cows in 53 German slaughterhouses and found that 9.6% of the female slaughter cattle were pregnant on average, and more than 90% of the affected animals were slaughtered during the last two trimesters of pregnancy. According to observations by the authors in several slaughterhouses in Latin American countries, it is not uncommon to find cows giving birth at slaughterhouses during lairage. The latter is probably related to the fact that many farmers delay culling cows until late lactation, when their yields become unprofitable, thus many are in a late stage of pregnancy. Hence, the delay in culling until the economically optimal time can also carry a welfare cost.

Efforts should be made to carefully inspect culled cows before loading and to coordinate each component of the pre-slaughter logistic chain to ensure the welfare of these animals (Miranda-de la Lama et al., 2014). Most regulations specify that animals should be fit for transport and that this must be checked before loading; hence there should be a better enforcement of the regulations. The welfare of animals can be improved by transporting and marketing culled breeding stock when they are still fit, before they become too weak, emaciated or even downer.

From an animal welfare perspective, it is extremely important to handle weak cows, and those with mastitis and lameness, carefully, because these animals will be in pain and distress, and this condition will follow or even aggravate during marketing, transport and slaughter (Fig. 1). These animals should not go through auction markets, and special conditions should be offered to them during transport (more space availability, bedding, etc.) in order to avoid further detrimental conditions during this period. If the cow is unfit for transport, a veterinarian should determine and supervise the implementation of the most appropriate killing method to ensure that animals are killed on farm without avoidable pain and distress (OIE, 2015). The recommended killing

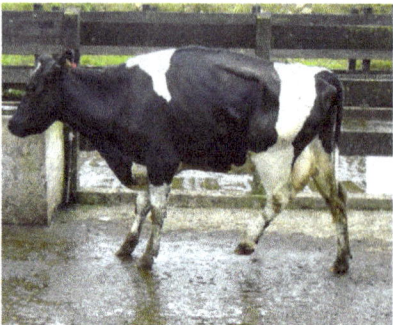

Figure 1 Culled cow with mastitis (left) and lame (right, photo by Dr N. Tadich) on a farm before being loaded onto the transport vehicle.

methods are free bullet, penetrating or non-penetrating captive bolt followed by bleeding, or injection with barbiturates and other drugs.

4 Welfare of culled cows during transport

Welfare during transport and slaughter can be assessed by behavioural and physiological measures, but also evaluation of injuries, bruises, mortality, morbidity and carcass quality can be used as indicators of welfare during handling and transport (Warriss, 1990; Knowles, 1999; Broom, 2000; Knowles and Warriss, 2007). Most of the studies on cattle transport have been done with steers and bulls destined for meat production and there are not many studies that have investigated the effects of transport conditions, duration of journey and others in culled cows. Already in 1978 Hails pointed out that the concern shown for an animal being transported increases in proportion to its economic value and culled cows have a low carcass value: lower dressing percentage, scarce or too much fat cover and poorer grades (Van Arendonk et al., 1984; Gallo et al., 1995, 1999; U.S., 2007). This fact, and the lower proportion of culled cows compared to other cattle categories supplying meat, probably explains why most experimental studies on the effects of transport have been done with cattle categories different from culled cows.

Transport reviews have focused mainly on the effects of transport on the animals such as mortality, behaviour, physiological responses and meat quality, and on environmental/external factors of the journeys, such as distance or time travelled, vehicle structure, handling, driver experience and others (Hails, 1978; Tarrant, 1990; Grandin, 1994; Knowles, 1999; Gallo and Tadich, 2005; Knowles and Warriss, 2007; Broom, 2008; Nielsen et al., 2011). Eicher (2001) refers more specifically to the transportation of cattle in the dairy industry and states that there is a need for additional research on the specific needs of different ages and stages of production of dairy cattle during transport, because transportation has become a routine management practice within dairy herds. Studies should include recovery of animals after transport and also consider the possible effects on productivity.

Grandin (2001) states that the most important issue in the case of transport is starting with an animal that is fit for the journey with serious problems occurring in culled breeding stock. There are only a few commercial studies dealing with the transport of culled dairy or beef cows and culled cattle in general (González et al., 2012a,b,c). According to González et al. (2012c), calves and culled cattle appear to be more affected by transport based on the likelihood of becoming lame, non-ambulatory or dead within a journey; in the same study also significantly more culled cattle were already lame at loading than other categories, particularly females. Therefore, culled cattle are at greatest risk of experiencing stress and poor welfare during long haul transport.

When culled cows are healthy and fit for the journey, most of the principles that apply for other cattle categories during transport will also apply to culled cows. However, this could be often not the case with culled cows transported in poor condition and/or because they are lame, with mastitis or other health problems, as basal values for stress indicators could be misleading.

4.1 Indicators of stress during transport

Even under good conditions of transport, cattle will show *physiological changes* that are indicative of stress (Broom, 2000). Increases in the concentrations of various blood components (cortisol, CK, packed cell volume, lactate, free fatty acids, betahydroxybutirate, total protein/albumin and others) and in physiological measures like heart rate, respiration rate and body temperature have been used as indicators of stress of animals during transport (Knowles and Warriss, 2007; Broom, 2008). Haptoglobin, a major acute phase protein, has also been used as an indicator of poor welfare during transport (Arthington et al., 2003). Research by Yagi et al. (2004) and Lomborg et al. (2008) in healthy dairy cows transported for 4 to 6 h evaluated the effect of short transport stress on somatic cell counts, migration of blood neutrophils and acute phase proteins in cattle and found that all these indicators increased after transport.

Regarding *behaviour during transport*, adult cattle normally prefer to stand and will not lie down in trucks while they are moving (Knowles, 1999); however at high stocking densities or with long journeys, they occasionally lie down (Tarrant, 1990). The most common standing orientations for cattle during long distance transport are perpendicular and parallel to the direction of travel, avoiding the diagonal orientations (Tarrant et al., 1992; Gallo et al., 2000, 2001). Maintenance of balance on moving vehicles requires energy and good footing, and more animals are likely to fall down after 12 h of transportation (Gallo et al.,

2000). The proportion of lying adult cattle increases with time after 12 h of transport, whereas young cattle (6-12 months) lie down earlier (Grandin and Gallo, 2007). When this happens at high stocking densities, cattle are trapped down and are unable to rise again (Tarrant et al., 1992).Transporting emaciated, lame or sick cows will render them more susceptible to loss of balance; moreover, when they fall during transport, it becomes particularly difficult for them to stand up again, and the risk of being trampled by other animals and suffocating becomes greater.

When an animal dies during transport, it is because its physiological mechanisms have failed to maintain homeostasis (Knowles and Warriss, 2007), and therefore, *mortality* is used to quantify stress. Adult cattle are more resilient than other livestock and mortality during transport appears to be low, but young calves are more vulnerable (Knowles, 1995, 1999). In dairy cows transported to slaughter between 1997 and 2004 in the Czech Republic, Vécerek et al. (2006) found that mortality rate during transport (died in the truck or shortly after unloading) was 0.038%. Mortality increased with travel distance, and season was also a factor in the death rates of dairy cows under transportation, with higher mortality occurring in the colder periods, as opposed to warmer periods. The same authors found a growing trend in dairy cow mortality during transport throughout the time studied, which is a warning sign in relation to the welfare of these animals. Malena et al.(2007) found a mean mortality rate of 0.0396 in dairy cows transported between 1997 and 2006 in the Czech Republic, with lowest mortality in less than 50 km distance transport (0.0137%) and highest (0.1874%) in transport longer than 300 km. In the United States, mortality rate recorded at arrival at the slaughterhouse was 0.04% (U.S., 2007).

The proportion of cattle *arriving as downers or non-ambulatory* can also be used as an indicator of poor transport. However, according to observations by the authors, in culled cows, this is commonly the result of the transport, plus the poor conditions of the animals before loading them (Fig. 2). Grandin and Gallo (2007) also indicate that some of the most severe welfare problems that occur during cattle transport are with culled cows and other animals that are sick, emaciated or debilitated, and hence of low economic value. Moreover, Grandin (1994) and Doonan et al. (2003) refer to non-ambulatory animals, which are mostly culled dairy cattle.

There are several *meat quality* problems – shrink, carcasses bruises, high ultimate pH – that become evident only after slaughter and reflect that animals have had a poor welfare during the pre-slaughter period (Warriss, 1990; Gallo and Huertas, 2016). Shrink is related to animal welfare as it reflects that animals have suffered from prolonged hunger and/or thirst. The loss of liveweight during long journeys causes shrink and can negatively affect carcass weight (Warriss, 1990). González et al. (2012b) showed that both feeder calves and culled cattle (bulls and cows) had the greatest shrink compared to calves and

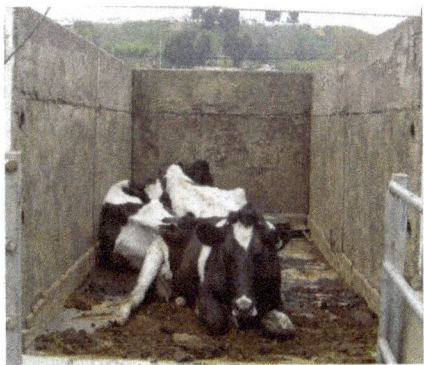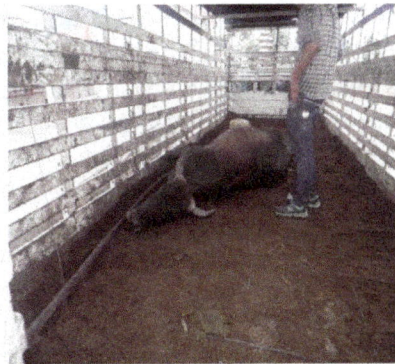

Figure 2 Culled cows arriving non-ambulatory at a Chilean (left) and Colombian (right, photo by Dr F. Ramírez) slaughterhouse.

fat cattle, in long distance commercial haul (>400 km, 12–16 h, without feed and water) in North America. Transport duration and cattle category were the most important factors affecting shrink.

The *presence of bruises* on a carcass is an indicator that cattle were exposed to improper handling and transport conditions and hence reflect a welfare problem during the pre-slaughter period. Bruising is also associated with economic losses due to trimming meat cuts and poorer meat quality. Older cows and oxen suffer more from bruising than younger cattle (Strappini et al., 2010). The presence of high ultimate pH (>6.0 at 24 h post-mortem) is an indicator that cattle have been through chronic stress during the pre-slaughter period (Ferguson and Warner, 2008), makes meat unattractive due to its darker colour and reduces its shelf life (DFD, Hood and Tarrant, 1980). In general, the incidence of high pH meat is higher in bulls and steers than in cows (Warriss, 1990). However, Knowles (1999) indicates that the high prevalence of DFD meat found in culled cows at slaughter within the United Kingdom suggests that they find marketing and transport to slaughter particularly stressful. This is consistent with the results of Strappini et al. (2010) who found that cows transported to the slaughterhouse, passing through the market, also have a higher percentage of high pH meat than those going directly from the farm, and that the presence of bruises was also significantly associated with increased carcass pH values. So, handling culled cows gently, reducing long fasting periods without water and food, and minimizing stress will undoubtedly help improve their welfare and also allow to produce more meat of better quality for the consumers.

4.2 Causes of animal stress during transport

There are many environmental/external factors related to cattle journeys that can increase stress and negatively affect animal welfare, and these are journey length, stocking density in the truck, vehicle design and maintenance, ventilation, climatic conditions, quality of roads and the standard of driving (Grandin and Gallo, 2007). The *time a journey takes* is generally more important than the distance travelled (Warriss, 1990) and there is often no direct relationship between them, particularly in countries with poor or geographically difficult roads, where a short distance could take many hours (Gallo and Tadich, 2008). Tarrant et al. (1992) found evidence of dehydration and fatigue after 24 h road transport and concluded that any extension of journey time or deterioration in transport conditions would be detrimental to the welfare of steers transported to slaughter. Warriss et al. (1995) concluded that a 15-h journey under good conditions of transport is not unacceptable from the viewpoint of animal welfare, based on the physiological measurements and subjective behaviour observations of castrated male cattle between 12 and 18 months of age. The findings of Gallo et al. (2000, 2001) who transported steers by road 3, 6, 12, 24 and 36 h (the latter with or without a rest stop) are similar to those of the above-mentioned authors, indicating that journeys of 24 h or more, without food and water, and at high stocking densities (500 kg/m^2) negatively affect welfare and meat quality.

A high *stocking density or low space availability* (1.05 m^2 per 600 kg steer) reduces the welfare of the cattle during transport and increases the stress response, number of falls and amount of bruising (Tarrant et al., 1992). Inappropriate space allowances can increase the frequency of falls, injuries, bruising, mortality, cortisol and creatine kinase concentrations (Tarrant et al., 1992; González et al., 2012c). González et al. (2012a) found that culled and breeding cattle in long distance hauling in North America had the fewest number of animals per trailer and hence the greatest space allowance (allometric coefficient k = 0.019-0.047, where k = space allowance/BW$^{0.6667}$) compared to other cattle categories; they suggest that more space is provided to animals with poorer body condition (culled) or high value (breeding). The same authors also found that gates in the deck were used more frequently when hauling calves, breeding and culled cattle, suggesting that this is due to the fact that culled and breeding cattle are usually loaded in groups of fewer animals and there is potential mixing of animals. This has also been observed in Chile by Gallo et al. (2005)and in South America in general by Gallo and Tadich (2008); these authors mention that in order to avoid welfare problems and deaths with the consequent economic losses for the farmers, culled cows are usually transported in short hauls, with higher space allowances, in smaller vehicles, and in smaller groups of animals to local slaughterhouses.

Nielsen et al. (2011) studied the effects of journey duration on animal welfare and concluded that transport of long duration is possible in terms of animal welfare provided that four issues can be dealt with, that are specific for the species and age group of the animals that are transported: the physiological and clinical state of the animal before and during transport; feeding and watering; rest and thermal environment. Hence, we consider that it would be possible to transport cattle for longer than 24 h provided that animals are healthy, have enough space to be able to rest (lie down), have comfortable bedding, have access to food and water, and other factors like ventilation, ambient temperature and humidity are controlled in accordance with the needs of the specific category of cattle transported (breed, age, sex, physiological state and others). The poor condition of culled dairy cows when leaving the farms, combined with transport stress, may become critical factors leading to impaired health and even to death during transport.

5 The effects of livestock markets on cow welfare

Cows can be transported for different reasons within farms, between farms, to the slaughter house when they are culled or to livestock markets to be sold (Fisher et al., 2008). This latter implies multiple loading/unloading, extra transports, and in some cases, mixing with unfamiliar animals.

According to González et al. (2012b), most of the culled cattle loads submitted to long haul transported in North America originate from auction markets (78%) and only a few from the farm/ranch (22%). Cattle loaded at auction markets are more likely to become non-ambulatory or die during transport compared to those loaded at farms or feed yards (González et al., 2012c). Something similar happens in South America, where much of the young fat cattle goes directly to the slaughterhouses, but feeder calves and culled cows often pass through livestock markets, where infrastructure and handling are poor (Gallo and Tadich, 2008). Several studies have shown that cattle passing through auction markets also suffer more bruising than those marketed directly from farm to slaughterhouse (McNally and Warriss, 1996; Weeks et al., 2002; Strappini et al., 2010) and that the most affected categories in terms of the number of bruises per carcass, as well as the size and depth of the lesions, are culled cows and oxen (Strappini et al., 2010, 2012).

De Vries (2011) describes the general design and structure of a livestock market as composed by an (un) loading point, races, holding pens, a weighing point and the arena or stage. When cows arrive at the market, they get out of the truck at the loading point. From this point they are driven through races to the holding pens. The size and design of the races, as well as the holding pens, can differ between markets. From the holding pens cows are driven again to the arena or selling yard. A stockperson in the ring (sometimes riding a horse) moves the animal to show the audience the qualities of the selling animal. The audience has the opportunity to bid, and the animal will be sold to the highest bidder. Just before (or after) being sold, the animals are weighed at the weighing point. From the weighing point or from the arena, cows are driven to a different holding pen or directly to the loading point where they are loaded on trucks and prepare to be transported again. The size of the animal group varies in relation to the design of the holding facilities. Mixing of unfamiliar animals in the holding pen can occur. Provision of water and food depends on market regulations, but in general it is not a common practice in many countries.

The movement of the cow from one stage to another inside the market might implicate stress, fear, fatigue and risk for injuries. In fact, studies on bruised carcasses from cows traded through livestock markets revealed several factors potentially harmful to the animals that impaired their welfare (Strappini et al., 2010). In this way, the gross characteristics of the bruises can be used to identify and evaluate potentially sub-optimal welfare conditions during the pre-slaughter period. The characteristics and number of bruises on Chilean cattle carcasses were studied and related to the source of the animals (Strappini et al., 2012). For a total of 258 cow carcasses (111 transported directly from farm to the slaughterhouse, and 147 cows traded via livestock market), the number of bruises, anatomical site, size, colour and shape were assessed. The number of bruises per carcass was higher in animals from markets than in off-farm animals (mean 3.8 ± 2.0 versus mean 2.5 ± 1.8 respectively). These results are in line with those reported by Jarvis et al. (1995) and Weeks et al. (2002) who found that 71.0% of the animals that had passed through a market showed a bruised carcass compared to 53.7% of the animals from farms. Regarding the distribution of bruises on carcasses, cows from markets showed more bruises on the hip, pin and ribs site presenting evidences of rough handling and animals beaten by sticks at markets (Strappini et al., 2012; Fig. 3).

Market animals in general showed agonistic behaviour and seemed to be excited displaying butting, mounting, defecation and vocalizations. Stockpersons of Chilean markets (De Vries, 2011) and other South American

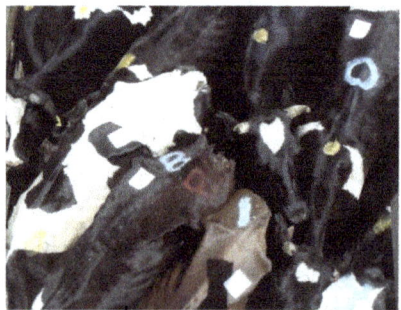 

Figure 3 Culled cows at a Chilean livestock market, showing horned cows of mixed origin at high stocking density in a pen (left) and cows being handled with sticks (right).

livestock markets (Gallo and Tadich, 2008) were observed using wooden sticks to poke and hit animals (Fig. 3). These findings are in line with Strappini et al. (2012) who found a high prevalence of bruises with a tram-line appearance reported on carcasses of cows coming from livestock markets in the same area of study (Fig. 4). These types of bruises are characteristic of those caused by a rounded shape object and are evidence that animals were beaten with sticks (Weeks et al., 2002).

The behaviour of the stockperson towards animals is likely to depend on the behaviour of their colleagues at the livestock market and it is influenced by subjective norms. However a busy environment might also have negative influence on the human – animal relationship since the stockperson feels pressure to move the animals quickly and does not take time to guide the animal with care. Therefore, the pressure exerted by a peer group as well as the working conditions can be the cause of the use of stick and rough handling (Coleman et al., 2000). Market authorities (i.e. staff, owners, official veterinarians) must ensure that cows exposed for sale at the livestock auction are in good condition and should have the authority to decide on the humane killing of those animals unfit for onward transport. Livestock markets operators should make it very clear that unfit animals will not be accepted in their markets. Therefore, farmers must be aware of the fact that unfit cows should not be transported, and the criteria to consider when an animal is unfit for transport are

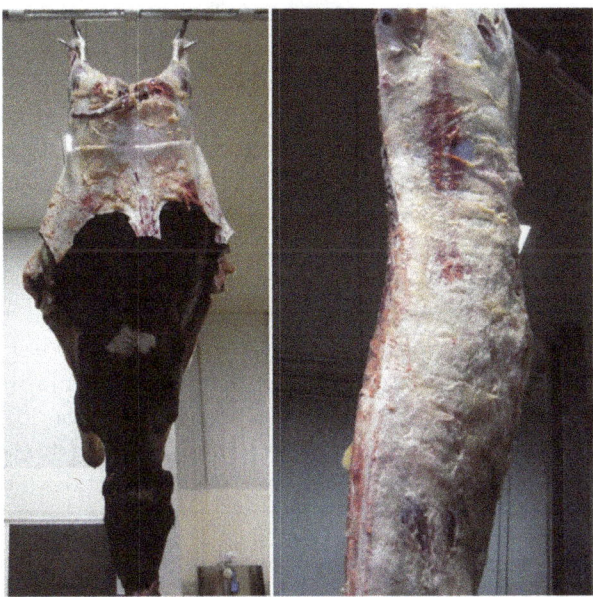

Figure 4 Stick marks that become visible after dehiding the cow at the slaughterhouse.

given by OIE (2015). Moreover, passing culled cows through auction markets should be avoided whenever possible as it will only add to further suffering and pain, and impair their welfare.

6 Welfare of cows at the slaughter plant

Grandin (1994) states that although culled cows represent less than 1% of the cattle handled in U.S. slaughter plants, they suffer greatly due to rough handling.

6.1 Arrival and unloading

When cattle arrive at a slaughterhouse, they are unloaded and then put in a lairage pen for ante-mortem inspection; the lairage pen also serves as a holding facility to organize cattle entering the slaughterline. Unloading should take place as soon as possible after arrival (OIE, 2015) in order to minimize further stress of the animals arriving tired from the journey. Regarding the arrival of cattle to the slaughterhouses, the longest unloading delays were found in culled cattle in North America (González et al., 2012b). In Colombia, Ramírez and Gallo (2012) observed that the mean waiting time for cattle before being unloaded was 6.3 h; there were frequently fallen animals in the trucks (Fig. 2), particularly weak cows, and the delay was due to the fact that cattle had to be unloaded only between 7.00 and 11.00 am at the slaughterhouse. These delays are common in other South American countries as well (Gallo and Tadich, 2008) and undoubtedly affect the welfare of the animals.

Non-ambulatory (downer) cattle are a major problem area and dairy cattle are 75% of the downers (Grandin, 1994). As observed by the authors, cattle that arrive as non-ambulatory in South American slaughterhouses are also mainly culled cows (Fig. 2). When unloading non-ambulatory cows, they should be handled and moved as little as possible in order to avoid producing more pain and distress on the animal. Dragging of downed or crippled livestock is forbidden in most countries. Slide boards and cripple carts are helpful to unload disabled cattle and further transport them to the emergency slaughterhouse to be immediately slaughtered. If this is not possible, the disabled animal should be stunned on the vehicle, before unloading, and slaughtered immediately after. However, it is still common to see improper unloading of non-ambulatory cows, being lifted or hoisted from different parts of their bodies instead of being humanely killed on the truck (Fig. 5).

6.2 Lairage duration and conditions

It has been well established that from an animal welfare perspective, as well as from a meat quality point of view, lairage time should be as short as possible

(Tadich et al., 2005; Ferguson and Warner, 2008). The OIE standards for animal welfare (2015) recommend that waiting times at the slaughterhouse be kept to a minimum and not be longer than 12 h; if animals need to be kept longer, food should be offered to them.

During lairage, cattle have access to water but not to food usually. The possibilities of resting are few. This may be due to the environment, which may not provide suitable conditions for rest, as there will be noise, unfamiliar smells and the presence of people and other animals. Gentle handling in well-designed facilities will minimize stress levels and improve efficiency at the slaughter plant; therefore, constant monitoring of handler performance and good maintenance of handling structures are required to maintain high standards of welfare (Grandin, 2007).

The authors have observed in several countries in South America that, with the exception of emergency slaughter, culled cows will usually remain in lairage for a longer period than other cattle categories, because preference for slaughter is given to steers and heifers that will render better carcasses and meat quality than culled cows. In Colombia, Ramírez and Gallo (2012) and Ramírez (2016) found that cattle in general remain between 12 and 118 h in lairage (mean 39.98 h, Fig. 6 left) and that culled cows experiment the longest lairage times, mostly with access to water but no feed. A similar feature was found by Gallo et al. (1995); with a mean of 34 h lairage for cattle in general, culled cows waited 39 to 51 h. Although lairage times are being reduced in most countries due to OIE (2015) recommendations and the negative effects on meat quantity and quality (Ferguson and Warner, 2008), in Chile, cattle still remain for over 12 h in lairage, as they usually arrive the evening before they will be slaughtered. This will affect their welfare in general, but particularly in the case of cows that are culled because of mastitis and lameness, which are painful and cows have to stay in pens without the necessary comfort. In fact, Meneses et al. (2005) examined 500 dairy cows at a slaughterhouse and found

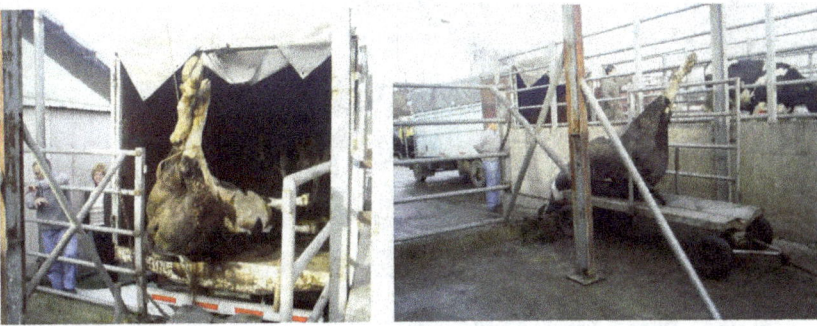

Figure 5 Cows being inadequately unloaded by lifting from their legs after arriving non-ambulatory at the slaughterhouse.

that 71% of them presented hoof lesions, not considering whether they were lame or not. The frequency of presentation of hoof lesions obtained in this study at a slaughterhouse is much greater than the farm prevalence of lame cows (Flor and Tadich, 2008; Hernández-Gotelli and Tadich, 2015); the higher prevalence of lame cows at slaughterhouses compared to farms shows that animal welfare issues in the culled cows concentrate at slaughterhouses.

Observations on the behaviour of cattle during lairage showed that the time taken for cattle arriving from markets to lie down after arrival at the slaughterhouse was less compared to those arriving directly from farms, and they were also lying down for a longer period (Cockram, 1991). Opitz et al. (2012) found that after short transport (100 km) during long periods of lairage (19 h) resting behaviour of culled cows (latency and duration of lying down) was uncommon, moreover a high number of interactions among them occurred (Fig. 6 right). Most cows remained standing during the whole period, only 24% of all cows were lying down at least once during the lairage period; the time they were lying down corresponded only to 5% of the lairage time. The physical condition also affected the behaviour of cows in lairage: cows in better condition interacted more and rested less. Hence, culled cows, particularly those that feel pain due to lameness or mastitis, will not rest and do not benefit from a long lairage.

6.3 Stunning and slaughter

Dairy cattle, cows in particular, have a close human–animal relationship and therefore a small flight zone which makes handling sometimes difficult when driving cows from the lairage pens to the stunning point at the slaughterhouse; as they are not afraid of people, they are difficult and slow to move. Strappini et al. (2013) found that after relatively short transport times (<4 h) directly from

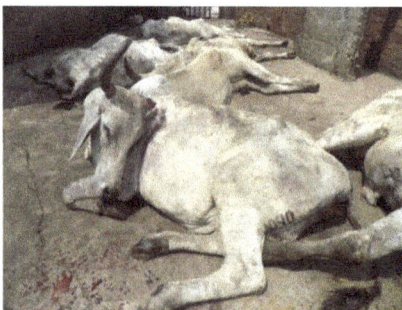

Figure 6 Dead and non-ambulatory cattle in lairage pens at a Colombian slaughterhouse (left, photo by Dr M. F. Ramírez) and culled cows during lairage at right, showing that they mainly remain standing.

farm to slaughterhouse and long lairage times (>12 h) most bruises in culled cows were the result of circumstances at the slaughterhouse; either due to human–animal, animal–animal or animal–facility interactions. One of the main problems was guillotine doors falling on the back of cattle. As cows are usually of larger size than steers and heifers, they are at greater risk of being hit with guillotine doors, when these doors are used to push cattle forward (Fig. 7).

The OIE (2015) recommends that cattle be immobilized before stunning and also that methods of restraint that cause avoidable suffering not be used. Grandin (2007) states that behavioural principles should be used for restraining animals, as this will enhance animal welfare and will reduce stress and injuries. Restraining devices include head gate designs, hydraulic chutes with adjustable sides, belly lifts and rear pushers. In all cases these devices should have enough pressure to provide the feeling of restraint, but avoid uncomfortable pressure points on the animal and pain on the animals. Immobilizing devices improve the efficiency of stunning (Gallo et al., 2003), but they can also produce stress if they exert excessive pressure on the animal (Ewbank et al., 1992). Muñoz et al. (2012) found that in 6.1% of the cases cattle restraint was incorrect, and this was significantly associated with vocalizations; cows were less affected by this, probably because of being more tame and familiar with holding structures on farm. However, large dairy cows were more affected than other cattle categories by the falling of guillotine doors on their backs.

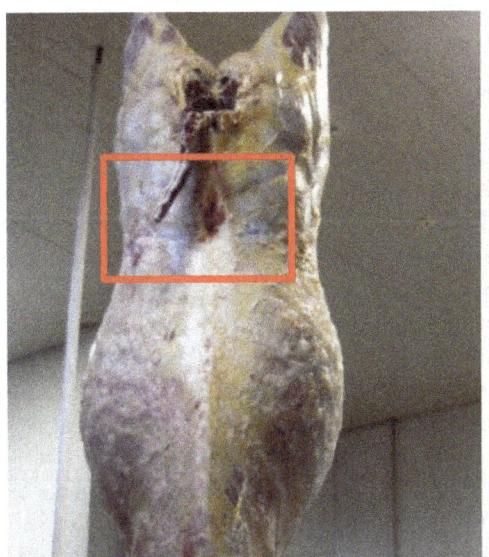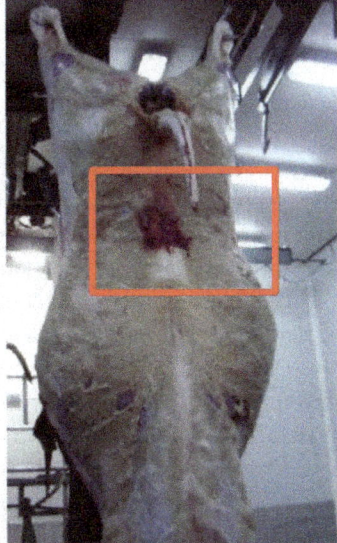

Figure 7 Carcasses of culled cows showing bruises, caused by falling of guillotine doors on their backs.

The most common method for stunning cattle is the captive bolt, which in its mode of action can be penetrating (which impacts the skull and enters the bolt into the brain) or non-penetrating (which only impacts the skull) (HSA, 2006). In order to avoid a possible return to consciousness, when using non-penetrating concussion stunners, animals should be bled within 30 s and it is not recommended for cattle of less than eight months of age, for mature stock bulls, or for aged cows (HSA, 2006). The OIE (2015) recommends that non-penetrating captive bolt be used only when there is no alternative method. The physical signs of an effective stun are: animal collapses, no rhythmic breathing, fixed/glazed expression in the eyes, no corneal reflex, relaxed jaw and tongue hanging out (HSA, 2006). Concha and Gallo (2009) compared the effectiveness of non-penetrating captive bolt between steers, heifers and cows during the stunning process, and found that cows had a higher presence of sensibility signs (rhythmic breathing, eye movements and corneal reflex, vocalizations and head lifting), the longest intervals between stunning and sticking, and also more frontal bone fractures, all indicators of incorrect stunning and impaired welfare. Stunning efficacy can be significantly improved by training stunning operators (Gallo et al., 2003).

Most of the welfare problems that culled cows suffer during stunning at the slaughter plant could be avoided by proper animal handling, adequate design of handling and stunning facilities and by training operators who understand that culled cows are sentient beings, independent of their lower economic value.

7 Conclusions

Culled cows should leave the farm before they become too weak or emaciated; emphasis must be on preventing cows to become non-ambulatory.

If cows become non-ambulatory on farm, they should be euthanized as soon as possible on site and not transported. If cows become non-ambulatory during transport or during lairage, they should be moved as little as possible so as to avoid further pain and slaughtered without delay.

Culled dairy cows should go directly from the farm to a slaughterhouse and not traded through cattle markets.

Duration of transport for culled dairy cows should be as short as possible and conditions of transport should consider providing a larger space availability than for steers and heifers, and comfortable bedding as to lie down.

Slaughterhouses should be organized in such a way to give preference to weak dairy cows, so that they can be humanely unloaded from the vehicles and slaughtered as soon as possible after arrival.

The training of people handling animals during transport and at the stunning point should incorporate the principle that all farm animals are sentient beings, regardless of the lower economic value of culled cows compared to other cattle categories.

There is need for more studies on the welfare of the different cattle categories during transport, particularly culled cows, and on the handling of them at the slaughterhouses, as there is some evidence that animal welfare during these procedures is greatly impaired in culled dairy cows.

8 Where to look for further information

For those who would like further reading on the welfare of culled dairy cows during transport and slaughter, and considering that information on this specific subject is scarce, we suggest to review the general legislation on transport and slaughter of animals for the countries of interest. We also recommend to read the OIE guidelines on animal welfare (OIE, 2015), particularly the standards on welfare of animals during transport and slaughter. For Spanish speaking readers we would also suggest the third edition of the book *Bienestar Animal: una visión global en Iberoamérica* (Mota-Rojas et al., 2016) which contains several chapters on good handling practices for cattle during the pre-slaughter period, including transport.

9 References

Arthington, J. D., Eicher, S. D., Kunkle, W. E. and Martin, E. G. 2003. Effect of transportation and commingling on the acute-phase protein response, growth and feed intake of newly weaned beef calves. *Journal of Animal Science* 81: 1120–5.

Australia. 2012. Land transport of livestock, Australian animal welfare standards and guidelines. Edition One, Version 1.1, 21 September 2012.

Bascom, S. S. and Young, A. J. 1998. A summary of the reasons why farmers culled cows. *Journal of Dairy Science* 81: 2299–305.

Broom, D. M. 2000. Welfare assessment and problem areas during handling and transport. In T. Grandin (ed.), *Livestock Handling and Transport*, 2nd edn., 43–61. Wallingford: CABI.

Broom, D. M. 2008. The welfare of livestock during road transport. In M. C. Appleby, V. Cussen, I. Garcés, I. Lambert and J. Turner (eds), *Long Distance Transport and Welfare of Farm Animals*, 157–81. Wallingford, UK: CABI.

Canada. 2015. Recommended code of practice for the care and handling of farm Animals Transportation. Health of Animals Regulations C.R.C., c296. http://laws-lois.justice.gc.ca.

Chile. 2013. Decreto N°30. Aprueba reglamento sobre protección del Ganado durante el transporte. Biblioteca del Congreso Nacional de Chile, Legislación Chilena. www.leychile.cl.

Cockram, M. S. 1991. Resting behavior of cattle in a slaughterhouse lairage. *British Veterinary Journal* 147: 109–19.

Coleman, G. J., McGregor, M., Hemsworth, P. H., Boyce, J. and Dowling, S. 2000. The relationship between beliefs, attitudes and observed behaviours of abattoir personnel in the pig industry. *Applied Animal Behavioural Science* 82: 189–200.

Concha, R. and Gallo, C. 2009. Evaluación del bienestar animal de bovinos durante el proceso de insensibilización con pistola de proyectil retenido no penetrante. XXXIV Congreso Anual de la Sociedad Chilena de Producción Animal, 21 al 23 de Octubre de 2009, Pucón, Chile. Libro Resúmenes, 260-1.

De Vries, M. 2011. Human-Animal Relationship at Chilean livestock markets. MSc. Thesis, Animal Science Department, Wageningen University, The Netherlands.

Doonan, G., Appelt, M. and Corbin, A. 2003. Nonambulatory livestock transport: The need for consensus. *Canadian Veterinary Journal* 44: 667-72.

Eicher, S. D. 2001. Transportation of cattle in the dairy industry: Current research and future directions. *Journal of Dairy Science* 84 (E Suppl): E19-23.

European Council. 2005. Council Regulation No 1/2005 of 22 December 2004 on the protection of animals during transport.

Ewbank, R., Parker, M. J. and Manson, C. W. 1992. Reactions of cattle to head-restraint at stunning: A practical dilemma. *Animal Welfare* 1: 55-63.

Farm Animal Welfare Council (FAWC). 2013. FAWC advice on space and headroom allowances for transport of farm animals.

Ferguson, D. M. and Warner, R. D. 2008. Have we underestimated the impact of pre-slaughter stress on meat quality in ruminants? *Meat Science* 80(1): 12-19.

Fetrow, J., Nordlund, K. V. and Norman, H. D. 2005. Invited review: Culling: nomenclature, definitions, and recommendations. *Journal of Dairy Science* 89: 1896-1905.

Fisher, A., Colditz, I., Lee, C. and Ferguson, D. 2008. The influence of land transport on animal welfarein extensive farming systems. *Journal of Veterinary Behaviour* 4: 157-62.

Flor, E. and Tadich. N. 2008. Lameness in cows from large and small dairy farms of Southern Chile. *Archivos de Medicina Veterinaria* 40: 125-34.

Gallo, C., Carmine, X., Correa, J. and Ernst, S. 1995. Análisis del tiempo de transporte y espera, destare y rendimiento de canal de bovinos transportados desde Osorno a Santiago. XX Reunión Anual de la Soc. Chilena de Producción Animal. SOCHIPA. Coquimbo, Chile.

Gallo, C., Caro, M., Villarroel, C. and Araya, P. 1999. Characteristics of cattle slaughtered within the Xth Region (Chile) according to the terms stated by the official Chilean standards for classification and carcass grading. *Archivos de Medicina Veterinaria* 31 (1): 81-8.

Gallo, C., Pérez, S. Sanhueza, C. and Gasic, J. 2000. Efectos del tiempo de transporte de novillos previo al faenamiento sobre el comportamiento, pérdidas de peso y algunas características de la canal. *Archivos de Medicina Veterinaria* 32 (2): 157-70.

Gallo, C., Espinoza, M. A. and Gasic, J. 2001. Efectos del transporte por camión durante 36 horas con y sin período de descanso sobre el peso vivo y algunos aspectos de calidad de carne en bovinos. *Archivos de Medicina Veterinaria* 33 (1): 43-53.

Gallo, C., Teuber, C., Cartes, M., Uribe, H. and Grandin, T. 2003. Mejoras en la insensibilización de bovinos con pistola neumática de proyectil retenido tras cambios de equipamiento y capacitación del personal. *Archivos de Medicina Veterinaria* 35 (2): 159-70.

Gallo, C., Warriss, P., Knowles, T., Negrón, R., Valdés, A. and Mencarini, I. 2005. Stocking densities used to transport cattle to slaughter in Chile. *Archivos de Medicina Veterinaria* 37 (2): 155-9.

Gallo, C. and Tadich, N. 2005. Transporte terrestre de bovinos: Efectos sobre el bienestar animal y la calidad de la carne. *Agro-Ciencia* 21 (2): 37-49.

Gallo, C. and Tadich, T. A. 2008. Chapter 10: South America. In M. C. Appleby, V. Cussen, L. Garcés, L. Lambert and J. Turner (eds), *Long Distance Transport and Welfare of Farm Animals*, 261-87. Wallingford, UK: CABI.

Gallo, C. B. and Huertas, S. M. 2016. Main animal welfare problems in ruminant livestock during preslaughter operations: A Southamerican view. *Animal* 10 (2): 342-8.

González, L. A., Schwarzkopf-Genswein, K. S., Bryan, M., Silasi, R. and Brown, F. 2012a. Space allowance during commercial long distance transport of cattle in North America. *Journal of Animal Science* 90: 3618-29.

González, L. A., Schwarzkopf-Genswein, K. S., Bryan, M., Silasi, R. and Brown, F. 2012b. Factors affecting body weight loss during commercial long haul transport of cattle in North America. *Journal of Animal Science* 90: 3630-9.

González, L. A., Schwarzkopf-Genswein, K. S., Bryan, M., Silasi, R. and Brown, F. 2012c. Relationships between transport conditions and welfare outcomes during commercial long haul transport of cattle in North America. *Journal of Animal Science* 90: 3640-51.

Grandin, T. 1994. Farm animal welfare during handling, transport, and slaughter. *JAVMA* 204 (3): 372-7.

Grandin, T. 1998. Handling of crippled and non ambulatory livestock. *Animal Welfare Information Center Bulletin* 9: 12-13.

Grandin, T. 2001. Perspectives on transportation issues: The importance of having physically fit cattle and pigs. *Journal of Animal Science* 79 (E Suppl.): E201-7.

Grandin, T. 2007. Handling and welfare of livestock in slaughter plants. Chapter 20. In T. Grandin (ed.), *Livestock Handling and Transport*, 3rd ed., 329-53. Wallingford, UK: CABI.

Grandin, T. and Gallo, C. 2007. Cattle transport. Chapter 9. In T. Grandin (ed.), *Livestock Handling and Transport*, 3rd ed., 134-54. Wallingford, UK: CABI.

Hails, M. R. 1978. Transport stress in animals, a review. *Animal Regulation Studies* 1: 289-343.

Hartung, J., Marahrens, M. and Holleben, K. V. 2003. Recommendations for future development in cattle transport in Europe. *Deutsche Tierârztliche Wochenschrift* 110: 81-132.

Hernández-Gotelli, C. and Tadich, N. 2015. Effect of lameness on culling of dairy cows in three herds in the South of Chile. Poster presented at the 18th International Symposium and 10th International Conference on Lameness in Ruminants, 22-25 November 2015, Valdivia, Chile.

Hood, D. and Tarrant, P. 1980. *The Problem of Dark Cutting in Beef*. The Hague, Netherlands: Martinus Nijhoff.

Humane Slaughter Association (HSA). 2006. Captive-bolt stunning of livestock. Guidance Notes N°2, 4th edition.

Jarvis, A. M., Selkirk, L. and Cockram, M. S. 1995. The influence of source, sex, class and pre-slaughter handling on the bruising of cattle at two slaughterhouses. *Livestock Production Science* 43: 215-24.

Knowles, T. G. 1995. A review of post-transport mortality among younger calves. *Veterinary Record* 137: 406–7.

Knowles, T. G. 1999. A review of the road transport of cattle. *Veterinary Record* 144: 197–201.

Knowles, T. G. and Warriss, P. D. 2007. Stress physiology of animals during transport. Chapter 19. In T. Grandin (ed.), *Livestock Handling and Transport*, 3rd ed., 312–28. Wallingford, UK: CABI.

Lomborg, S. R., Nielsen, L. R., Heegaard, P. M. H. and Jacobsen, S. 2008. Acute phase proteins in cattle after exposure to complex stress. *Veterinary Research Communications* 32: 575–82.

Malena, M., Voslarova, E., Kozak, A.,Belobradek, P., Bedanova,I., Steinhauser, I. and Vécerek, V. 2007. Comparison of mortality rates in different categories of pigs and cattle during transport for slaughter. *Acta Veterinaria Brno* 76: S109–16.

McNally, P. W. and Warriss, P. D. 1996. Recent bruising in cattle at abattoirs. *Veterinary Record* 138: 126–8.

Meneses, E., Baez, A. and Tadich, N. 2005. Lesiones podales en vacas destinadas a matadero. XXX Reunión Anual de la Sociedad Chilena de Producción Animal. 19 al 21 de Octubre, Temuco, Chile.

Miranda-de la Lama, G. C., Villarroel, M. and María, G. A. 2014. Livestock transport from the perspective of the pre-slaughter logistic chain: A review. *Meat Science* 98: 9–20.

Mota-Rojas, D., Velarde, A., Huertas, S. M. and Cajiao, M. N. 2016. Bienestar Animal, Una visión global en Iberoamérica. Third edition, Elsevier, Barcelona, España.

Muñoz, D., Strappini, A. and Gallo, C. 2012. Animal welfare indicators to detect problems in the cattle stunning box. *Archivos de Medicina Veterinaria* 44: 297–302.

New Zealand. 2011. Animal Welfare (Transport within New Zealand) Code of Welfare 2011.

Nielsen, B. L., Dybkjaer, L. and Herskin, M. S. 2011. Road transport of farm animals: Effect of journey duration on animal welfare. *Animal* 5 (3): 415–27.

OIE. World Organization for Animal Health. 2015. Terrestrial Animal Health Code, Section 7. Chapter 7.3. Transport of animals by land and Chapter 7.5. Slaughter of animals.

Opitz, C., Strappini, A., Vargas, R. and Gallo, C. 2012. Análisis descriptivo de las conductas realizadas por las vacas durante el período de espera en corrales de una planta faenadora. 17° Congreso Chileno de Medicina Veterinaria, Valdivia, Chile, 18–20 Noviembre de 2012.

Ramírez, M. F. 2016. Estudio sobre el manejo de bovinos en una planta faenadora en Colombia y sus efectos sobre el bienestar animal y la calidad de la carne. Tesis Magíster en Ciencias, mención Salud Animal, Facultad de Ciencias Veterinarias, Universidad Austral de Chile, Chile.

Ramírez, M. F. and Gallo, C. 2012. Evaluación preliminar del bienestar animal en una planta de faenamiento en Colombia. 17° Congreso Chileno de Medicina Veterinaria, Valdivia, Chile, 18–20 Noviembre de 2012.

Randall, J. M. 1993. Environmental parameters necessary to define comfort for pigs, cattle and sheep in livestock transporters. *Animal Production* 57: 299–307.

Riehn, K., Domel, G., Einspanier, A., Gottschalk, J., Lochmann, G., Hildebrandt, G., Luy, J. and Lücker, E. 2014. Slaughter of pregnant cattle: Aspects related to ethics and animal welfare. In L. Mounier and I. Veissier (eds), *Proceedings of the 6th International*

Conference on the Assessment of Animal Welfare at Farm and Group Level, 256. Clermont_Ferrand, France, 3-5 September 2014.

Strappini, A., Frankena, K., Metz, J. H. M., Gallo, C. and Kemp, B. 2010. Prevalence and risk factors for bruises in Chilean bovine carcasses. *Meat Science* 86: 859–64.

Strappini, A. C., Frankena, K., Metz, J. H. M., Gallo, C. and Kemp, B. 2012. Characteristics of bruises in carcasses of cows sourced from farms or from livestock markets. *Animal* 6: 502-9.

Strappini, A. C., Metz, J. H. M., Gallo, C., Frankena, K., Vargas, R., De Freslon, I. and Kemp, B. 2013. Bruises in culled cows: When, where and how are they inflicted. *Animal* 7(3): 485-91.

Tadich, N., Gallo, C., Bustamante, H., Schwerter, M. and van Schaik, G. 2005. Effects of transport and lairage time on some blood constituents of Friesian cross steers in Chile. *Livestock Production Science* 93: 223-33.

Tarrant, P. V. 1990. Transportation of cattle by road. *Applied Animal Behaviour Science* 28: 153-70.

Tarrant P. V., Kenny, F. L., Harrington, D. and Murphy, M. 1992. Long distance transportation of steers to slaughter: Effect of stocking density on physiology, behavior and carcass quality. *Livestock Production Science* 30: 223-38.

U.S. 2007. Cow and bull beef quality audit. Chapter 5. Quality assurance of market cows and bulls, pp. 22-4.

U.S.A. 2003. Code of Federal Regulations. 9.Parts 1 to 199. Animals and animal products.

Van Arendonk, J. A. M., Stokvisch, P. E., Korver, S. and Oldenbroek, J. K. 1984. Factors determining the carcass value of culled dairy cows. *Livestock Production Science* 11: 391-400.

Vécerek, V., Malena, M. Jr., Malena, M., Voslarova, E. and Bedanova, I. 2006. Mortality in dairy cows transported to slaughter as affected by travel distance and seasonality. *Acta Veterinaria Brno* 75: 449-54.

Warriss, P. D. 1990. The handling of cattle pre-slaughter and its effects on carcass and meat quality. *Applied Animal Behaviour Science* 28: 171-86.|

Warriss, P. D., Brown, S. N., Knowles, T. G., Kestin, S. C., Edwards, J. E., Dolan, S. K. and Phillips, A. J. 1995. Effects on cattle of transport by road for up to 15 hours. *Veterinary Record* 136: 319-23.

Weeks, C. A., McNally, P. W. and Warriss, P. D. 2002. Influence of the design of facilities at auction markets and animal handling procedures on bruising in cattle. *Veterinary Record* 150: 743-8.

Yagi, Y., Shiono, H., Chikayama, Y., Ohnuma, A., Nakamura, I. and Yayou, K. 2004. Transport stress increases somatic cell counts in milk and enhances the migration capacity of peripheral blood neutrophils of dairy cows. *Journal of Veterinary Medical Science* 66 (4): 381-7.

Chapter 3

Optimising pig welfare during transport, lairage and slaughter

Luigi Faucitano, Agriculture and Agri-Food Canada, Canada; and Antonio Velarde, Institute of Agrifood Research and Technology, Spain

1 Introduction

Stress imposed on pigs during transport, in lairage and at slaughter is both an animal welfare and a meat quality issue. Studies revealed that poor handling practices and transport conditions and wrong facility design at the slaughter plant both, individually and/or additively, can contribute to the loss of profits due to animal losses (dead-on-arrival or DOA and downers), reduction in carcass value due to weight losses and skin bruises and meat quality defects due to abnormal *post-mortem* muscle acidification (Schwartzkopf-Genswein et al., 2012; Faucitano, 2018; Rioja-Lang et al., 2019).

The responsibility for animal losses during transport may either be equally shared among the producer, who must guarantee proper preparation of animals (i.e. feed withdrawal) prior to transport and handling of pigs at loading, the trucker and the abattoir (in the case of integrated production systems) or be bore alone by the trucker, whereas the abattoir is responsible alone for the optimisation of lairage and slaughter conditions (layout, ambient control and handling systems) in order to let pigs recover from the stress of handling and transport and ensure optimal and uniform carcass and meat quality.

Training of handling crews and the application of rewards and fines have become of paramount importance to improve handling and reduce animal

http://dx.doi.org/10.19103/AS.2020.0081.07

losses (Correa, 2011; Rocha et al., 2016; Dalla Costa et al., 2019). A Canadian integrated company reported a decrease from 0.3% to 0.01% in the incidence of fatigued pigs on arrival at the abattoir by applying a new programme of payment with incentives to handlers and truckers as a reward for their willingness to slow down the loading/unloading speed rate to 100 pigs/h (Correa, 2011). In contrast, fines up to $6000 under Canadian Food Inspection Agency regulations are applied to Canadian truckers for having –three to four dead pigs in the truck load (Faucitano, 2018). Grandin (2018) recently reported that the decision of an abattoir management in the US to apply a $25 handling fee for each pig arriving in a non-ambulatory condition resulted in a great reduction in the occurrence of downers.

This chapter will focus on the handling practices to be applied during the transportation of pigs (market weight and culled sows) to slaughter, as this phase is considered as the most stressful of the whole preslaughter period, and in lairage and at stunning/slaughter at the abattoir, aiming at limiting their impact on animal losses and pork quality.

2 Welfare during transport

The situation when a pig is in transit is considered a major stressor and may have deleterious effects on its health, its well-being, carcass yield and ultimately pork quality. Major sources of stress during transport are, among others, vehicle design, the time spent in the truck and the space allowed to lie down.

2.1 Vehicle design

Vehicle design features that may influence the welfare of pigs during transport include the loading/unloading system (ramps or hydraulic decks), the compartment location and the microclimate control (Faucitano and Goumon, 2018; Rioja-Lang et al., 2019).

The presence of fixed decks and ramps within the vehicle makes the procedures of loading and unloading more difficult, resulting in a higher risk of animal losses (Barton-Gade et al., 2007; Correa et al., 2013) and poor pork quality, that is, either pale, soft and exudative (PSE) or pale or dark, firm and dry (DFD) pork (Guàrdia et al., 2004; Correa et al., 2013, 2014). The North American pot-belly (PB) trailer is an example of the vehicle design featuring two to five, often steep (up to 32° slope), ramps (see Fig. 1) and 180° turns that pigs have to negotiate during the loading and unloading process. These trailer features have been associated with a lower ease of handling as showed by the greater use of electric prods and longer handling time (Ritter et al., 2008; Torrey et al., 2013a,b; Weschenfelder et al., 2013b), eventually resulting in a greater proportion of DOA and fatigued pigs on arrival at the abattoir compared with other trailer models featuring hydraulic decks (Ritter et al., 2008; Sutherland

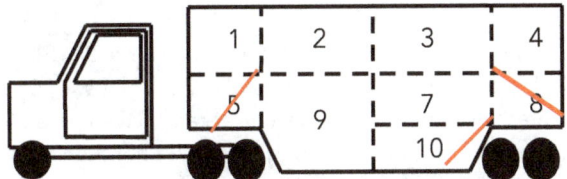

Figure 1 Position of compartments and internal ramps (solid red lines) inside a pot-belly trailer (modified from Correa et al., 2014). The represented model is a dual-purpose pot-belly trailer (for cattle and pig transport).

et al., 2009; Kephart et al., 2010; Weschenfelder et al., 2012; Correa et al., 2013).

The presence of ramps is one of the major contributors to the handling problems reported during loading and unloading in PB trailers (see Figs. 1 and 2; Faucitano and Goumon, 2018; Rioja-Lang et al., 2019). It is well known, in fact, that negotiating a ramp slope from 0° to 45° represents a significant physical and psychological challenging experience for pigs, as showed by the increased heart rate, frequency of turn backs and baulking, skin bruises and

Figure 2 Unloading from a pot-belly trailer through an internal ramp (L. Faucitano, AAFC, Canada).

Published by Burleigh Dodds Science Publishing Limited, 2021.

handling time (Van Putten and Elshof, 1978; Warriss et al., 1991; Dalla Costa et al., 2007; Goumon et al., 2013b; Garcia and McGlone, 2015). Recently, Dalla Costa et al. (2019) compared ramp slopes of < 20° versus > 20° and reported a four-fold greater risk for DOA when steeper ramps were used. Furthermore, steep ramps (up to 26°) are also challenging for the handler, as showed by the increased heart rate and difficulty to handle (Goumon et al., 2013b). Current guidelines do not recommend to use ramps steeper than 20° for fixed ramps or than 25° for adjustable ramps (SCAHAW, 2011; TQA, 2016). However, as market pigs have become heavier (from 113 kg to 130 kg from 2000 to 2010 in Canada; Correa, 2011) and difficult to move (Bertol et al., 2011; Rocha et al., 2016), recent works recommend the maximum ramp slope to be reduced to 15° (Grandin, 2012; Goumon and Faucitano, 2017).

2.2 Animal position in the truck

The impact of the deck and/or compartment position on the vehicle pigs' welfare during transport and meat quality can be explained by the differences in the ease of handling (due to the presence of ramps), airflow rate, vibration rate and loading order (Faucitano and Goumon, 2018). European and Canadian studies showed that, when compared with the middle deck, pigs transported in the top and/or bottom decks presented higher body temperature, exsanguination blood cortisol levels, dehydration rates, carcass lesion score and incidence of PSE and exudative or pale/dark pork (Lambooij et al., 1985; Lambooij and Engel, 1991; Barton-Gade et al., 1996; Correa et al., 2013). The front and rear top and bottom rear compartments can be also critical, as showed by the increased core body temperatures after loading and during transport and higher heart rate after loading and longer unloading time (Goumon et al., 2013a; Torrey et al., 2013b; Conte et al., 2015). A greater risk for DFD pork production has been reported in pigs transported in the middle front compartment of a PB trailer (Correa et al., 2014), especially after long transportation (18 vs. 6 and 12 h) in winter (Scheeren et al., 2014).

2.3 Control of the microclimate within the transport vehicle

Generally, greater animal losses are reported during summer hauls (Vecerek et al., 2006; Werner et al., 2007; Haley et al., 2008; Correa et al., 2013; Vitali et al., 2014), with greater risk being recorded at ambient temperatures either between 17°C and 20°C (Warriss and Brown, 1994; Sutherland et al., 2009; Haley et al., 2008, 2010; Kephart et al., 2010) or between 29°C and 33°C (Peterson et al., 2017).

Within a vehicle the internal temperature can increase by almost 1°C for each 1°C increase in the environmental temperature (Dewey et al., 2009). The internal temperature increases more, up to 3-4°C rise in 5 min at a rate of approximately 1°C/min, in passively ventilated vehicles held in a stationary situation (Lambooij, 1988; Xiong et al., 2015). However, when passively ventilated trailers are compared, the environment is warmer in the PB trailer compared with the triple-deck flat-deck (FD) trailer while stationary or moving (Ritter et al., 2008; Weschenfelder et al., 2012; Faucitano and Goumon, 2018). The variation in the internal thermal conditions between the two trailer models has been explained by the difference in the side openings type (punch for the PB vs. slatted for the FD trailer) influencing the air flow inside the vehicle (Weschenfelder et al., 2012). More specifically, within the PB trailer the highest temperature peak (30.3°C) has been recorded in the bottom deck during transport (Brown et al., 2011), while the greater gradient (up to +11°C) between internal trailer and external ambient temperatures has been recorded in the middle and bottom front compartments during stops (Weschenfelder et al., 2012, 2013; Fox et al., 2014).

To ensure thermal comfort and reduce animal losses during vehicle stops, pigs should be cooled off by fan-assisted ventilation, water sprinkling/misting or the two systems combined. Research showed that when applied for 5 min after loading and before leaving the farm and at the end of the wait before unloading at the abattoir at ambient temperatures of 20°C and greater, water sprinkling reduced pigs' fatigue (as assessed by the blood lactate concentration) at slaughter and pork exudation, especially in pigs transported in the middle front and rear compartments (Nannoni et al., 2014). However, water sprinkling combined with insufficient ventilation can also result in an increased difference in humidity levels (up to +7.5%) between the trailer interior and the external environment (Fox et al., 2014), preventing efficient evaporative cooling in pigs. Pereira et al. (2018) showed that the use of fan-misting bank installed in the abattoir yard (see Fig. 3) for 30 min under warm ambient conditions (up to 28.1°C) was efficient to improve the quality of the internal environment of a PB trailer (lower temperature–humidity index or THI) and the pigs' thermal comfort (lower body heat loss) while waiting before unloading.

A greater proportion of DOAs and non-ambulatory pigs on arrival at the plant has been also reported in winter than in summer (Guàrdia et al., 1996; Sutherland et al., 2009), with a higher risk of death at 4–10°C than at 12–26°C (Peterson et al., 2017). Likely causes of the greater animal losses in winter compared with other seasons may be the more difficult animal handling through the icy internal ramps at loading and unloading (Torrey et al., 2013a,b) and insufficient bedding of the trailer floor, resulting in more pigs standing during transport to avoid the contact with the cold aluminium floor surface (Goumon et al., 2013a). The presence of slippery ramps can result in more slips and falls at loading and unloading (Torrey et al., 2013b), greater heart rates during transport and unloading (Goumon et al., 2013a) and increased blood creatine kinase (CK) and lactate concentrations at slaughter (Correa et al., 2014).

Figure 3 Operation of a fan-misting bank on pigs kept in a pot-belly trailer during their wait before unloading (L. Faucitano, AAFC, Canada).

In winter, the thermal comfort of pigs in the truck can be controlled by partially closing the ventilation openings (boarding) in order to reduce air flow and by adding 5-cm layer of Styrofoam to the vehicle top ceiling (Gonyou and Brown, 2012) and bedding on the truck floor (TQA, 2016). Within the air temperature range between 15°C and ≤ −12°C, trucks should be utilising from 25% to 95% boarding to prevent DOAs (McGlone et al., 2014a; TQA, 2016).

In winter, no more than six bales of bedding should be sufficient to avoid pigs to be in contact with the cold floor surface of the trailer during transport and reduce the rate of DOA or downers and the number of carcass lesions, including frostbites (Goumon et al., 2013a; McGlone et al., 2014b; Scheeren et al., 2014). The number of bales should be reduced to three in summer (McGlone et al., 2014b).

2.4 Journey time

Both long and short journeys can affect the welfare of slaughter pigs and mortality rate during transport (Rioja-Lang et al., 2019). Some studies reported an increased risk of animal and body weight losses (Warriss et al., 1990; Sutherland et al., 2009), fatigue and dehydration, as shown by the higher blood glucose, lactate and haematocrit levels at slaughter (Brown et al., 1999; Mota-Rojas et al., 2006; Becerril-Herrera et al., 2010; Sommavilla et al., 2017), and DFD pork production (Leheska et al., 2003; Mota-Rojas et al., 2006) with transport time longer than 4 h and up to 24 h. In contrast, in other studies shorter journeys (≤2 h) resulted in a lower ease of handling at the abattoir, greater concentrations of cortisol and lactate in exsanguination blood and a higher risk of PSE pork production (Faucitano and Lambooij, 2019). A possible reason for the detrimental effect of short journeys is that the pigs could not recover from the stress of handling experienced at loading upon arrival at the abattoir (Weschenfelder et al., 2013).

However, the effects of travel time on pig welfare can be additive to that of other concurrent factors, such as the time on fasting, season of the year, type of vehicle, loading density and pig genetic background (Rioja-Lang et al., 2019). Fasting pigs on farm for 12-18 h before loading reduces the risk of mortality during transport, regardless of the journey duration (up to 24 h), while in unfasted pigs this risk increases for journeys up to 8 h (Averós et al., 2008). In summer increased risk for DOA has been reported in hauls longer than 2 h (Vitali et al., 2014), while greater core body temperature, heart rate and incidence of DFD pork production have been recorded in pigs transported for 18 h in winter than in summer (Goumon et al., 2013a; Scheeren et al., 2014). The latter results may be caused by the longer exposure of pigs to cold stress. In transportation trials of different durations (45 min and 7 h), using both PB and FD trailers, Weschenfelder et al. (2013) reported an increased level of fatigue (based on exsanguination blood lactate concentration) at the time of slaughter in stress-susceptible pigs (Hal[Nn]) hauled for a shorter time with the PB trailer compared with a FD trailer equipped with semi-hydraulic middle and top decks.

2.5 Space allowance

The recommended loading density for pigs during transport involves a trade-off between the economic pressure to increase the loading density in order to maximise profit from a single journey and the welfare of animals during transport (Rioja-Lang et al., 2019). However, particular attention must be paid to this transport variable considering its significant contribution to animal losses. In a US survey of more than 12 000 loads, Fitzgerald et al. (2009) reported that trailer density accounted for the largest portion of the variation in the total animal losses, that is, dead, fatigued and injured pigs, compared to other variables, such as the driver, handling crew, THI, wind speed, loading duration and wait time before unloading.

The EU legislation (Council Regulation (EC) No 1099/2009, 2009) is based on the evidence that when the loading density is higher than 235 kg/m^2 or lower than 0.425 m^2/100 kg not all pigs are able to lie down to rest and cannot rest as they are pushed to continually change their position (Lambooij et al., 1985; Lambooij, 2014). This uncomfortable situation has been associated with increased mortality rates and a higher number of non-ambulatory pigs on arrival at the plant (Ritter et al., 2006). Fitzgerald et al. (2009) reported that increasing the density from 212.4 kg/m^2 to 338.6 kg/m^2 corresponded to a 7.5-fold increase in animal losses on arrival at the slaughter plant. However, both low (0.50 m^2/100 kg) and high density (0.33 m^2/100 kg) during transport can cause physical stress, resulting in muscle fatigue and glycogen depletion, and a greater risk of DFD pork production (Lambooij et al., 1985; Guàrdia et al., 2005). In a low-density situation, physical stress is caused by the attempt of pigs to maintain their balance and cope with the unexpected and sudden movements of the truck or by fighting due to their greater freedom to move around in the compartment (Barton-Gade and Christensen, 1998; Guàrdia et al., 2005). In contrast, providing less space may fatigue pigs due to the frequent disturbance of lying animals by those seeking a place to rest and the difficulty of standing pigs to maintain their balance while the vehicle is moving (Lambooij and Engel, 1991).

The application of loading densities should be adjusted according to the body weight, ambient conditions and travel time. A recent Italian study reported a greater risk of DOAs when the EU-recommended density (235 kg/m^2) is applied for the transport of 160 kg pigs, which would correspond to a loading density of 0.7 pigs/m^2 (Nannoni et al., 2016). Because of their different physical needs and greater susceptibility to heat stress, due to the decreased heat dissipation rate (Renaudeau et al., 2011), the minimum recommended truck space required by heavier-market-weight pigs (from 114 kg to 182 kg) should be increased from 0.40 m^2/pig to 0.61 m^2/pig in winter and from 0.46 m^2/pig to 0.65 m^2/pig in summer (Grandin, 2017b). If the load size is simply determined by the number of animals that need to be shipped, then heavier pigs may be too packed on the truck (Zurbrigg et al., 2017). During hauls of 140-kg weight pigs, the reduction of the number of pigs from 3/m^2 to 2/m^2 has been associated with a decreased risk of animal losses from 1% to 0.2% (Fitzgerald et al., 2009).

During hot and humid ambient conditions, it is recommended to provide 15% to 25% more space (CARC, 2001; https://ec.europa.eu/food/animals/w elfare/practice/transport_en). Indeed, transport losses are lower when more truck floor space (0.46 m^2/100 kg vs. 0.39 m^2/100 kg) is provided in summer (Ritter et al., 2012). However, greater space allowance (0.50 m^2/100 kg vs. 0.25 m^2/100 kg) may increase the risk for skin lesions (+28.2%) in this season (Guàrdia et al., 2009). In winter, the risk of skin lesions is higher (9.7%) at high loading density (0.25 m^2/100 kg) due to the effect of huddling to keep the body temperature and fighting/mounting to seek for space to lie down (Guàrdia et al., 2009).

Pilcher et al. (2011) showed that the increase of floor space (from 0.40–0.49 m^2/100 kg to 0.52 m^2/100 kg) helps reduce the incidence of fatigued pigs on arrival at the plant after short transport (<1 h) compared with longer journeys (3 h) as pigs can get adapted more to a long journey (550 km) as showed by the higher proportion of pigs lying on the truck floor and reduced body temperature and heart rate (Gerritzen et al., 2013). However, short journeys (1 h vs. 3 h) at higher loading densities (0.25 m^2/100 kg vs. 0.50 m^2/100 kg) have been also associated with a decreased risk of PSE pork production (Guàrdia et al., 2004). Based on this evidence, in order to prevent this meat quality defect, the EU-recommended space allowance of 0.425 m²/100 kg should be only applied for journeys longer than 3 h. These results may be explained by the fact that at the departure from the farm pigs do not lie down immediately but still stand exploring and getting adapted to the novel environment. They remain so for the first few kilometres (0.5 h) to better cope with the fear and stress (as showed by the increased heart rate up to 220 beats/min) caused by the frequent truck accelerations/decelerations and vibrations generated by driving on the initial country unpaved roads (Chevillon, 2001). It can take up to 2 h of transport for the pigs to settle down in the truck (Lambooij et al., 1985; Barton-Gade and Christensen, 1998). In these conditions, providing pigs with less space would help them keep their balance by holding each other while the vehicle negotiates bends or poor road surfaces (Barton-Gade and Christensen, 1998).

2.6 Transport of culled sows

In pig production, sows are most likely culled due to poor body condition, lameness or failure to rebreed (Baloyh et al., 2015; Zhao et al., 2015; Grandin, 2016a). Such defects, along with greater mortality, are mostly reported at the largest farms (Koketsu, 2000).

Because of poor body and ambulatory conditions, cull sows may hardly walk and be loaded onto the truck, resulting in a greater risk of arriving fatigued, lame and even with worst body conditions, or dead, at the destination (i.e. buying station or slaughter plant) compared with market-weight pigs (Malena et al., 2007; McGee et al., 2016; Peterson et al., 2017). In a recent observational study on 47 loads of sows transported from the farm to slaughter, Thodberg et al. (2019) observed a deterioration of clinical signs, such as vulva and udder ulcers, skin lesions, wounds and torn hoofs, on arrival at the slaughter plant. These observations support the recommendation for producers to select fit sows and boars for transportation (Bench et al., 2008; Grandin et al., 2016a).

The effects of transportation on sow welfare can be exacerbated by the very long distance cull sows have to travel from the farm or assembly yard to the abattoir and ambient conditions. A recent study surveying the cull sow market network in the US reported that sows could be marketed between 24 and 120 h and travel straight to slaughter for an average of 1057 km prior to harvest (Blair and Lowe, 2019). Peterson et al. (2017) reported that for cull sows the risk to die during transport may be 1.93 and 0.81 times higher at outdoor ambient temperatures ranging from 29°C to 33°C and from 4°C to 10°C, respectively, compared to 12°C to 26°C.

Transport duration (4 h on average, ranging from <1 to 8 h), temperature in the truck (14°C on average, ranging from 3°C to 26°C) and duration of stops and wait before unloading at the slaughter plant (0.5 h on average for both, ranging from 0 h to 3 h and from 0 min to 75 min, respectively) have also been identified as the most important causing factors for the deterioration of clinical signs in cull sows on arrival at the abattoir (Thodberg et al., 2019).

3 Welfare in lairage

The purpose of lairage is to give stressed or fatigued animals an opportunity to recover from the stress of transport and previous handling in order to produce better meat quality (Faucitano, 2010, 2018; Gallo et al., 2016).

The recovery rate of pigs in lairage and the related economic losses due to poor carcass and meat quality depend on lairage time, the quality of the handling systems and ambient control (Faucitano, 2010; Gallo et al., 2016).

3.1 Lairage time

Two to three hours is usually the recommended rest time to allow the pigs to recover their physiological condition after transport and handling and to ensure the production of good-quality pork (Warriss, 2003). Unless required by the adversive ambient conditions, such as too high ambient temperature (>30°C; Fraqueza et al., 1998) or ammonia levels (>10 ppm; Weeks, 2008), short lairage intervals (<1 h) should not be applied as they may result in a higher muscle temperature and lactate level at slaughter, resulting in an increased incidence of PSE pork production (Fraqueza et al., 1998; Shen et al., 2006). However, while longer lairage time helps reduce the risk of PSE pork production (Guàrdia et al., 2005), extending lairage time from 3 h to overnight increases the risk of DFD pork production (Guàrdia et al., 2005) due to the depletion of the muscle glycogen content at slaughter caused by the combined effect of fasting and fighting within mixed groups of unfamiliar pigs (Nanni Costa et al., 2002; Guàrdia et al., 2009; Dalla Costa et al., 2016). Fighting-related skin lesion scores, in fact, increase with the lairage time (Warriss, 1996; Faucitano, 2001, 2010; Bottaccini et al., 2018), with almost a two-fold higher risk in pigs kept in lairage for 15 h compared to 3 h (Guàrdia et al., 2009).

3.2 Mixing

Under commercial conditions, mixing groups of unfamiliar pigs in lairage is a common practice because of the difference in size between the truck compartment and lairage pen, despite the risk of fighting, in terms of biting, pushing and head knocks (Rabaste et al., 2007) between pigs, and carcass value downgrading due to severe skin lesions (Faucitano, 2001). However, some strategies can be applied to limit fighting in lairage, such as keeping pigs in smaller groups (10 pigs/group vs. 30 pigs/group; Rabaste et al., 2007) and reducing the space allowance (0.85 m²/pig to 0.42 m²/pig; Moss, 1978; Geverink et al., 1996; Weeks, 2008). Anecdotally, some European plants managed to reduce fighting by enriching the pen with some corn kernels scattered on the floor that may distract pigs or sprinkling the back of pigs with vinegar that apparently mask the individual smell of each pig (Eyes on Animals, 2019).

The intensity and duration of fighting in lairage can be also aggravated by pig gender, with boars being more aggressive than gilts and barrows (Warriss and Brown, 1985) and immunocastrates fighting two-fold more than surgical castrates, especially when fed ractopamine during the last 28 days of the finishing period (Rocha et al., 2013). In a more recent study, Wesoly et al. (2015) reported an inverse relationship between testicular function and skin lesion score within mixed groups of boars, with high-ranking boars presenting lower skin lesions compared with low-ranked ones that are mounted more frequently (Rydhmer et al., 2006).

3.3 Ambient control

The control of temperatures between 15°C and 18°C and relative humidities (RH) between 59% and 65% is recommended to ensure the thermal comfort of pigs in lairage (Honkavaara, 1989). When these environmental conditions are not respected, pigs can either suffer from cold stress (shivering and huddling), which may result in DFD meat due to muscle energy depletion to maintain a constant body temperature (Knowles et al., 1998) or heat stress, as showed by increased panting, especially when they are kept at hot (>30°C) and humid (RH > 80%) conditions (Santos et al., 1997). According to Grandin (2012), death losses in lairage, also called 'dead-in-pen', almost doubled at temperatures above 32°C compared with 16°C.

Water spraying on pigs in the lairage pen helps reduce the heat-related respiration rate (Huynh et al., 2006), the number of dead-in-pen (Vitali et al., 2014) and the incidence of PSE pork production through a 2°C drop in the muscle mass temperature (Long and Tarrant, 1990). However, at temperatures below 5°C, showering is not recommended as it causes animal shivering and may lead to DFD pork due to muscle energy depletion to maintain a constant body temperature (Knowles et al., 1998). Ivn order to remove the excessive humidity produced by the application of sprinkling/misting systems as well as to control the concentration of noxious gases, for example, ammonia (Weeks, 2008), this practice must be combined with a proper ventilation (135 m^2 h^{-1}; Brent, 1986; Weeks, 2008). Vitali et al. (2014) reported a lower risk of death in lairages characterised, among others, by efficient ventilation through large open windows on the roof and side walls.

When compared to other livestock species (i.e. cattle and sheep), pig lairages are the loudest ones (Weeks et al., 2009), with the average noise level ranging from 76 dB to 108 dB and the highest peaks (120 dB) being recorded in the *peri-mortem* area (Talling et al., 1996; Rabaste et al., 2007). Excessive lairage noise produces a fear response in pigs, as showed by the number of pigs huddling in the pen looking for protection or escaping from the source of sound (Geverink et al., 1998), the increased heart rate and greater blood lactate, CK and cortisol levels at slaughter (Faucitano, 2010), all resulting in an increased production of PSE pork (Warriss et al., 1994; Van de Perre et al., 2010). Keeping the sound level lower than 85 dB in the *peri-mortem* area appears to reduce the risk for PSE meat (Vermeulen et al., 2015).

Lairage noise is mostly caused by gates clanging, operating machinery, echoes and pig vocalisation (Weeks, 2008; Weeks et al., 2009), although pigs appear to be more stressed by industrial sounds than the sounds of conspecifics (Geverink et al., 1998). To reduce the ambient sound level, some European plants replaced metal gates and fencing with plastic ones and modified the ceiling by decreasing the height and installing sound-absorbing materials (Eyes on Animals, 2019).

The movement of pigs in the alleys of the slaughter plant may be influenced by the lighting of these areas. Pigs are less reluctant to move from a dark area to a brightly lighted area (Van Putten and Elshof, 1978; Grandin, 1990; Tanida et al., 1996), making the use of moving tools less necessary. Grandin (2010) observed that the insufficient light (less than 160–215 lux) at the entrance of the stunning area increased the use of electric prods by 34%. The alleys must be consistently lighted to avoid shadows and contrasts in colour on the floor. It has been recently observed that the use of green lighting reduces shadows on the floor and improves the ease of handling through the alleys (Eyes on Animals, 2019; see Fig. 4).

3.4 Driving pigs to slaughter

The combination of a higher slaughter speed, poorly designed handling systems and the progressive passage from a free-moving group situation to a single line of aligned and restrained individuals during the short period of time between the exit from the lairage pen and stunning may result in a greater proportion of slips, jamming, backing-up and vocalisation (Warriss et al., 1994; Edwards et al., 2010, 2011; Van de Perre et al., 2010; Vermeulen et al., 2015; Rocha et al., 2016), and increased use of electric prods (Rocha et al., 2016). These behavioural responses have been associated with increased heart rate (up to 240 beats/min; Chevillon, 2001; Correa et al., 2010), blood lactate and CK levels at slaughter (Hambrecht et al., 2005; Edwards et al., 2010; Rocha et al., 2015), skin lesions scores (Rabaste et al., 2007) and exudative pork (Van der Wal et al., 1999; Hambrecht et al., 2005; Rabaste et al., 2007; Dokmanović et al., 2014; Rocha et al., 2016).

In the *peri-mortem* area critical factors are the entrance into the stun chute and the 'stop-start' forward movement of pigs towards the stunner,

Figure 4 Green lighting reduces the presence of shadows on the alley floor preventing stops and backups slowing down the flow of pigs towards slaughter (Photo courtesy of Eyes on Animals, The Netherlands).

which are both observed in raceways feeding electrical and CO_2 stunners and which result in the frequent use of electric prods to encourage the pig movement (Faucitano, 2010). The batch flow system or curved raceway has been observed to work well, easing the flow of pigs into the single/double electric stun raceway, as showed by the lower number of reluctant-to-move-forward pigs and the use of electric prods compared with the single file chute entrance (Edwards et al., 2010). However, this system must be well designed and managed, in that it must not have an abrupt entrance and should not be filled with pigs for more than 75% of its capacity (ideally 50%) to prevent jamming and overlaps in the crowd pen (Grandin, 2017b; see Fig. 5). The entrance of pigs into a CO_2 gas stunner has been significantly improved by the group-wise stunning system (Christensen and Barton-Gade, 1997), where pigs are moved forwards in groups of 15 pigs using pushing gates and are loaded in sub-groups of 5 into the cradle. This system proved to reduce the frequency of PSE (−3%) and blood-splashed pork (−5%) and bruised carcasses (−4%) due to reduced electric prodding and physical stress (Christensen and Barton-Gade, 1997).

4 Welfare during stunning and slaughter

The last stage of pig production is the slaughter for human consumption. The slaughter process includes the restraining of the animal, the stunning application and the exsanguination.

4.1 Restrainer types

The purpose of restraint is to facilitate the correct application of the stunning and bleeding. The restrainer type varies depending of the stunning method. In electrical stunning, restraint is needed to facilitate the correct placement of the electrodes between the eye and the ear and to facilitate the uninterrupted flow of current. However, incorrect restraint may lead to ineffective stunning or bleeding but can also cause pain and distress in its own right. The restrainer should match the size of the pigs. It should be narrow enough that the animals cannot move backwards and forwards or turn around. Insecure restraint may cause struggling or escape attempts (Grandin, 2016b). On the other hand, a too narrow device will provide excessive force to the animal and will be painful and may cause injuries. Prolonged restraint time may cause fear and exacerbate insecure or excessive restraint. Therefore, no pig should enter the restrainer until equipment and personnel are ready to slaughter that animal.

Several automatic electrical stunning methods are currently available. One device consists of a V-type restrainer where each pig makes contact with the electrodes and receives the stunning current (Fig. 6). Both sides must run at the

Figure 5 Filling the crowd pen with pigs for more than 75% of its capacity prevents easy handling into the stunning chute (L. Faucitano, AAFC, Canada).

Published by Burleigh Dodds Science Publishing Limited, 2021.

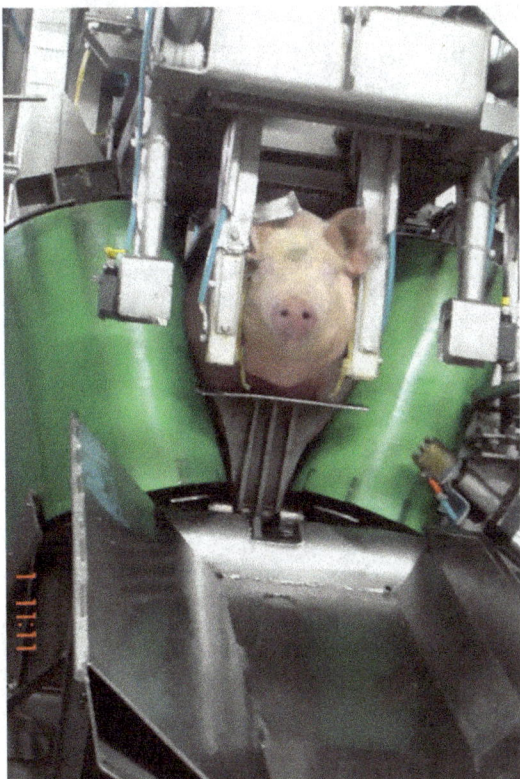

Figure 6 V-type restrainer (A. Velarde, IRTA, Spain).

same speed (Grandin, 2016b). When the two conveyors run at different speeds and their angle is too steep, the animals may struggle. A second method uses a conveyor belt system (Fig. 7). In this system, at the end of the restrainer, the nose of the pigs interrupts a beam of light, which activates the electrodes. In both systems, the animals are turned out and fall onto a table after the stunning.

Animal-based measures to assess pain, fear or distress of the pigs in the restrainer include slips or falls, struggling, escape attempts and vocalisation (Velarde and Dalmau, 2012; Grandin, 2016b).

Comparative studies showed that pigs restrained in the conveyor belt presented a lower heart rate (180 heartbeats/min vs. 220 heartbeats/min) and proportion of PSE pork production compared with those conveyed with the V-type restrainer (Griot et al., 2000).

On the other hand, the exposure to CO_2 at a high concentration does not require restraint as animals are stunned in groups (Velarde et al., 2000). In this case, gondolas for the gas stunning of pigs should not be overloaded, and animals should be able to stand without being on top of each other. Gondolas

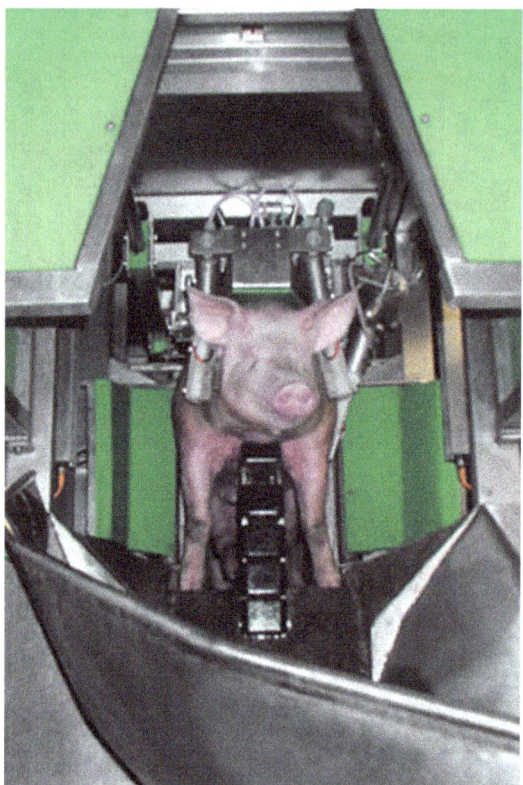

Figure 7 Conveyor belt system (A. Velarde, IRTA, Spain).

overloading is a criterion in animal welfare auditing protocols at slaughter plants (Grandin, 2017b) based on the observation of 25-50 gondolas, depending on the plant size. Audit scores are given on a gondola basis and range from excellent (no gondolas are overloaded) to serious problem (the person moving the animals forces one or more pigs to jump on top of the other pigs in the gondolas with an electric prod or by hitting, shoving or kicking).

4.2 Stunning methods

To spare any avoidable pain, distress or suffering, stunning before bleeding is a common practice around the world. Stunning is defined as any intentionally induced process which causes the loss of consciousness and sensibility without pain, including any process resulting in instantaneous death (Regulation No. 1099/2009). During the loss of consciousness, the animal is unable to perceive external stimuli and control its voluntary mobility and, therefore, does not respond to normal stimuli, including pain (EFSA, 2004). In order to fulfil the humane slaughter requirements, the duration of unconsciousness must be longer than the sum of time that lapses between the end of stun and the time to the onset of death due to bleeding. Therefore, unconsciousness should be monitored at different stages of the process: (1) during shackling and hoisting, (2) before sticking and (3) during bleeding (EFSA, 2013). The EFSA AHAW panel has developed toolboxes of welfare indicators for developing monitoring procedures at slaughterhouses for pigs (EFSA, 2013). The most commonly used methods for stunning pigs at slaughter are electrical stunning and exposure to CO_2 at high concentrations.

Electrical stunning (Fig. 7) involves the application of an electric current of sufficient magnitude to the brain such that a generalised epileptiform activity is induced similar to that recorded in humans during grand mal epileptic seizures (Croft, 1952; Hoenderken, 1978). This seizure-like state, immediately followed by an exhausted state, is suggestive of an immediate loss of consciousness and appears to be associated with a lack of sensory awareness, which lasts a finite period of time (Anil, 1991). An effective stun is characterised by the presence of all of tonic–clonic seizures, the loss of posture, apnoea and the absence of corneal reflex (Velarde and Dalmau, 2012). The main hazards preventing effective stunning are incorrect electrode placement, poor contact, dirty or corroded electrode, too low voltage/current or high frequency (EFSA, 2004).

Exposure to CO_2 at high concentrations is also used for the stunning of pigs. In this case, the loss of consciousness is not immediate (Rodríguez et al., 2008). Since CO_2 does not induce immediate loss of consciousness, inhalation of concentrations greater than 30% of CO_2 by volume in the atmospheric air causes aversion, irritation of the mucous membranes (that can be painful) and respiratory distress during the induction phase (Velarde et al., 2007). During CO_2 exposure, pigs show signs of aversion such as retreat attempts, headshaking, sneezing, breathlessness, freezing, escape attempts, gasping (a very deep breath through a gasping-open mouth, indicative of breathlessness; Raj and Gregory, 1996) and vocalisations (Holst, 2001; Velarde et al., 2007).

The main hazards causing increased distress during induction to unconsciousness are irritant or aversive gas mixtures, low gas temperature or humidity.

The exposure to anoxic gases (argon or nitrogen) with less than 2% by volume of residual oxygen in air is non-aversive and does not cause sense of breathlessness before the loss of consciousness occurs. However, the time to induce unconsciousness when exposed to anoxia is longer than when exposed to hypercapnia (Raj et al., 1997) and might not be commercially feasible. Raj and Gregory (1996) reported that the addition of up to 30% CO_2 to an anoxic atmosphere reduces the time needed to induce unconsciousness with minimal aversion. Argon (Ar) has a low presence in the atmosphere (0.9% by volume), and its availability for commercial stunning practices might be limited. On the other hand, the relative density of nitrogen (N_2) is slightly lower than air concentrations and cannot be sustained within a pit at a higher concentration than 94% by volume (Dalmau et al., 2010). Nevertheless, this stability could be improved when nitrogen and CO_2 are combined (Dalmau et al., 2010).

The EU Regulation (EC 1099/2009) has approved the low atmospheric pressure stunning (LAPS) for broiler chickens. It consists of the exposure to gradual decompression with reduction in the available oxygen to less than 5% (Martin et al., 2016). Current research is carried out in pigs to assess the effectiveness in inducing unconsciousness without aversion.

An effective stun is characterised by the presence of the loss of posture, apnoea, absence of corneal reflex and absence of muscle tone (Velarde et al., 2007). The main hazards causing ineffective stunning are incorrect gas concentration or short gas exposure time (EFSA, 2004). In case of ineffective stunning or recovery, re-stun immediately using a backup system taking into account the causes of failure or recovery.

Comparative studies reported the positive effects of CO_2 stunning on the pork quality when compared with electrical stunning (Velarde et al., 2000, 2001). However, the effects of gas stunning on animal welfare and pork quality depend on the gas concentration and exposure time. Better results, in terms of the percentage of clinical reflexes and risk for PSE pork, were obtained after stunning with a higher CO_2 concentration (90% vs. 80%) and longer gas exposure time (100 vs. 70 sec; Nowak et al., 2007).

4.3 Exsanguination techniques

Exsanguination consists of the severance of the arteries supplying oxygenated blood to the brain. Pigs are bled by chest sticking, severing the brachiocephalic trunk and the major blood vessels which arise from the heart (Fig. 8). With an adequate incision, pigs lose between 40% and 60% of their total blood volume, and within 30 sec, the amount of blood lost is between 70% and 80% of the total amount of blood which will be lost (Warriss and Wilkins, 1987). The time to lose brain responsiveness (based on the reduction in visual evoked responses) ranges between 14 sec and 23 sec (mean 18 ± 63) and the development of an isoelectric electrocorticogram (ECoG) between 22 sec and 30 sec (Wotton and Gregory, 1986).

The main welfare concern at the time of bleeding following stunning is the recovery of consciousness due to prolonged stun-to-stick interval or due to incomplete severance of the main blood vessels.

Published by Burleigh Dodds Science Publishing Limited, 2021.

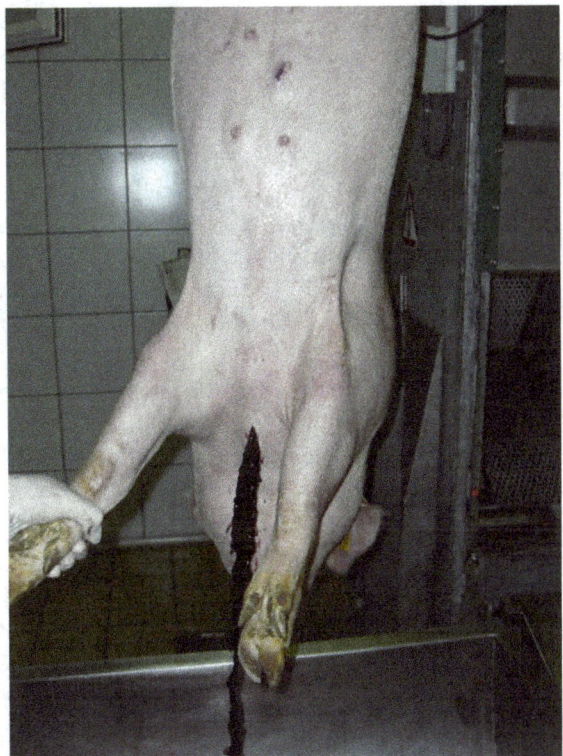

Figure 8 Chest sticking (A. Velarde, IRTA, Spain).

5 Animal welfare audit protocols

Nowadays, there is an increasing need for credible assessment systems to determine the welfare status of animals prior to slaughter. The application of animal welfare auditing protocols allows the evaluation of handling and slaughter practices, resulting in a significant improvement in handling practices, facilities design and quality of work, besides reassuring concerned consumers and increasing market opportunities due to the availability of certified products (Ballantyne, 2006; Grandin, 2017a).

Because of these benefits, animal welfare auditing has been increasingly implemented during transport and at slaughter since 1999 in the US, Australia, New Zealand and Europe, and in Canada since 2001 (Grandin and Smith, 2004).

Presently, the two main protocols used to assess animal welfare during transport and lairage and at slaughter are the audit protocols developed by Temple Grandin for the North American Meat Institute (NAMI; Grandin, 2017b) and by researchers of the European project Welfare Quality® (WQ®, 2009).

The NAMI audit protocol is based on the 'Recommended animal handling guidelines and audit guide' and has the objective to evaluate animal welfare during transport and at the slaughter plant through the assessment of seven core criteria based on the observation of animal behaviour (falls and vocalisation), assessment of facilities design, electric prod use, wilful acts of abuse, efficiency of stunning systems (electrical and CO_2 gas), access to water and insensibility on the bleed rail (Grandin, 2017b). The respect of the NAMI guidelines at the slaughter plant is a requirement to comply with the standards of the Humane Farm Animal Care and for the pork product to get the Certified Humane Raised and Handled® label (Humane Farm Animal Care, 2018).

The WQ° protocol (WQ°, 2009) has developed an integrated and standardised welfare assessment system based on 12 welfare criteria grouped into four main principles (good feeding, good housing, good health and appropriate behaviour) according to how they are experienced by animals (Dalmau et al., 2009). The innovation of the WQ° assessment system is its greater focus on animal-based measures (e.g. body condition, health, injuries and behaviour) rather than on resource and management-based measures (e.g. space allowance, number of drinkers and truck management). The WQ° welfare assessment starts in the unloading area, where general fear, thermoregulation behaviours, slipping and falling, sickness and dead animals are observed (Velarde and Dalmau, 2012). In lairage, five criteria are taken into consideration, namely the absence of thirst, hunger and disease; thermal comfort; and comfort while resting. The absence of thirst is calculated based on the number of drinking points in each pen and drinker functionality and cleanliness. The availability of feed for animals that have been held more than 12 h in the holding pen is the measure for the absence of hunger, while the absence of disease is measured by the presence of dead animals in the pen. Thermal comfort is evaluated by behavioural thermoregulation measures, such as huddling, shivering or panting scores. The last, but not least, criterion is comfort while resting in the pen, which is evaluated based on the space allowance. During movement from the lairage pen to the stunning area, good human–animal relationship is assessed by recording the incidence of high-pitched vocalisations. Afterwards, stunning effectiveness is assessed by observing the signs of the absence of pain, such as corneal and righting reflex, breathing and vocalisation, immediately after stunning and before sticking. After slaughter, skin lesion scoring provides valuable information regarding the management of animals on the farm of origin, during transport or in the lairage pen. Furthermore, the health status of the animals on the farm of origin is also assessed after slaughter by inspecting the internal organs for the presence of pleurisy and pneumonia in the lungs, pericarditis in the heart and white spots in the liver.

The efficiency of the WQ® programme as a tool for animal welfare monitoring has been scientifically validated at the abattoir (Dalmau et al., 2009). In 2014, a certification scheme of animal welfare based on the WQ® protocols has been developed by the Animal Welfare Subprogram of the IRTA (Monells, Spain) and the Spanish Association for Standardisation and Certification (AENOR). In 2017, already several pig abattoirs had been certified. The final objective of this programme is to include the certificate in the pork meat product label, named as 'Animal Welfare. AENOR Conform' (Dalmau, 2017). Recently, the IRTA has become scheme owner for the Animal Welfare Certification 'based on Welfare Quality®'.

However, neither the WQ® nor the NAMI assessment protocols cover the production process as a whole (from farm to slaughter). Rocha et al. (2016) developed a novel protocol for the animal welfare assessment of the whole pork chain by merging the WQ protocol with those of the Canadian Animal Care Assessment (CPC, 2011), which is a more resource- and management-based measurement of animal welfare at the farm level (including loading), and that of the American Meat Institute (AMI; Grandin, 2012) and validated it under Canadian commercial conditions. The objective of this study was to assess the relationship between the audit scores obtained at animal welfare-improved (AWI) and conventional farms using the WQ and CPC protocols and the variation in pig behaviours at loading, at unloading and in lairage as assessed by the WQ and AMI audit protocols. In this study, the frequency of slips used as an animal welfare audit criterion in the AMI protocol at the slaughter plant showed to be a good predictor of drip loss variation in the loin muscle, with muscle exudation being significantly related to slips during unloading ($r = 0.63$) and immediately before stunning ($r = 0.74$).

6 Conclusion and future trends

This chapter overviewed the effects of the stressors experienced by pigs during the transport, lairage and slaughter on animal losses; behavioural and physiological responses to stress; and carcass and meat quality.

Research provided the evidence that the use of truck models featuring hydraulic ramps or decks can help reduce the workload of handlers and improve the welfare of pigs during transport. However, more research on the truck design is needed with a study of the air flow patterns and vibration rate and insulation and cooling systems under different ambient conditions where temperature control becomes more critical and physiological heat maintenance and dissipation in pigs of different market weights become less effective. So far, most swine transportation research focused on slaughter weight pigs (up to 130 kg), but there is a considerable lack of information for the transportation of heavier pigs, including cull sows and boars. More specifically, information on their thermal and physical needs during transport and relative science-based guidelines for space allowance and travel time are needed.

Lairage and slaughter are extremely important for the pork chain economy as mistakes made at these points have irreversible effects on the welfare of pigs and carcass and meat quality and may offset all efforts made by the production sector to improve performance and animal welfare. Precautions must be taken to ensure proper handling and ambient control to keep the benefits of lairage as a resting area, allowing pigs to recover from the stress of transport and previous handling. The correct management and monitoring of critical areas in lairage is becoming paramount in the light of the increasing need for commercial abattoirs to obtain animal welfare audit approval and certification for their meat products.

7 Where to look for further information

7.1 Further reading

- Comprehensive and up-to-date reviews on swine transportation and animal welfare are the chapters authored by Faucitano and Goumon (2018) and Faucitano and Lambooij (2019) and published in the books *Advances in Pig Welfare* and *Livestock Transport and Handling*.
- An excellent source of science-based guidelines for the best handling practices at the abattoir (from reception to slaughter) in multiple species is the book *Animal Welfare at Slaughter: a Practitioner Guide* by Raj and Velarde (2016).
- Guidelines and recommendations on preslaughter handling and animal welfare can be found in the Temple Grandin web page: http://www.grandin.com/.
- A complete introduction to the relationship between animal welfare preslaughter and meat science is represented by the books *Animal Welfare and Meat Science* (1998) and *Animal Welfare and Meat Production* (2007) by Neville Gregory.
- A catalogue of training handbooks, guidelines and technical sheets in relation to transport and slaughter are available in the web page of the Humane Slaughter Association: https://www.hsa.org.uk/.

7.2 Key journals/conferences

- Special issues on pig transportation in the *Animals* journal (2014/2016): https://www.mdpi.com/journal/animals/special_issues/pig-trans-2016; https://www.mdpi.com/journal/animals/special_issues/pig-trans.
- Proceedings of special sessions on preslaughter handling and animal welfare included in the programme of the annual congresses of the

International Society of Applied Ethology (ISAE; https://www.applied-ethology.org/), the International Congress of Meat Science & Technology (ICoMST; http://www.icomst.helsinki.fi/index.htm) and American and Canadian Societies of Animal Science (ASAS-CSAS) Meeting & Trade Show; https://www.asas.org).

7.3 Major international research projects

- The Canadian Agricultural Partnership (http://www.agr.gc.ca/eng /about-us/key-departmental-initiatives/canadian-agricultural-part nership/?id=1461767369849) and Swine Innovation Porc (http://www.swineinnovationporc.ca/) are supporting projects on swine transportation.

8 References

Anil, M. H. (1991), 'Studies on the return of physical reflexes in pigs following electrical stunning', *Meat Sci.*, 30, 13–21.

Averós, X., Knowles, T. G., Brown, S. N., Warriss, P. D. and Gosálvez, L. F. (2008), 'Factors affecting the mortality of pigs being transported to slaughter', *Vet. Rec.*, 163, 386–90.

Ballantyne, W. (2006), 'A proactive approach to animal welfare', *Adv. Pork Prod.*, 17, 139–44.

Baloyh, P., Kapelankis, W., Jankowiak, H., Nagy, L., Kovacs, S., et al. (2015), 'The productive lifetime of sows on two farms and reasons for culling', *Ann. Anim. Sci.*, 15, 747–58.

Barton-Gade, P. and Christensen, L. (1998), 'Effect of different loading density during transport on welfare and meat quality in Danish slaughter pigs', *Meat Sci.*, 48, 237–47.

Barton-Gade, P., Christensen, L., Brown, S. N. and Warriss, P. D. (1996), 'Effect of tier and ventilation during transport on blood parameters and meat quality in slaughter pigs', *EU-Seminar: New Information on Welfare and Meat Quality of Pigs as Related to Handling, Transport and Lairage Conditions*, Landbauforschung Völkenrode, Kulmbach, Germany, 1996, vol. 166, 101–16.

Barton-Gade, P., Christensen, L., Baltzer, M. and Petersen, L. (2007), 'Causes of pre-slaughter mortality in Danish slaughter pigs with special emphasis on transport', *Anim. Welf.*, 16, 459–70.

Becerril-Herrera, M. M., Alonso-Spilsbury, M., Trujillo-Ortega, M. E., Guerrero-Legarreta, I., Ramirez-Necoechea, R., et al. (2010), 'Changes in blood constituents of swine transported for 8 or 16 h to an abattoir', *Meat Sci.*, 86, 945–8.

Bench, C., Schaefer, A. L. and Faucitano, L. (2008), 'The welfare of pigs during transport', in L. Faucitano and A. L. Schaefer (Eds), *Welfare of Pigs: From Birth to Slaughter*, Wageningen, The Netherlands: Wageningen Academic Publishers, pp. 161–95.

Bertol, T. M., Ellis, M., Ritter, M. J. and McKeith, F. K. (2011), 'Effect of feed withdrawal and handling intensity on longissimus muscle glycolytic potential and blood measurements in slaughter weight pigs', *J. Anim. Sci.*, 83, 1536–42.

Blair, B. and Lowe, J. (2019), 'Describing the cull sow market network in the US: a pilot project', *Prev. Vet. Med.*, 162, 107–9.

Bottaccini, M., Scollo, A., Edwards, S., Contiero, B., Veloci, M., et al. (2018), 'Skin lesion monitoring at slaughter on heavy pigs (170 kg): welfare indicators and ham defects', *PLoS ONE* (open access). doi: 10.1371/journal.pone.0207115.

Brent, G. (1986), *Housing the Pig*, Ipswich, UK: Farming Press Ltd.

Brown, S. N., Knowles, T. G., Edwards, J. E. and Warriss, P. D. (1999), 'Relationship between food deprivation before transport and aggression in pigs held in lairage before slaughter', *Vet. Rec.*, 145, 630–34.

Brown, J., Samarakone, T. S., Crowe, T., Bergeron, R., Widowski, T. M., et al. (2011), 'Temperature and humidity conditions in trucks transporting pigs in two seasons in eastern and western Canada', *Trans. ASABE*, 54, 1–8.

CARC. (2001), 'Recommended code of practice for the care and handling of farm animals – transportation', Canadian Agri-Food Research Council, Ottawa, Canada, 63p.

Chevillon, P. (2001), 'Pig welfare during pre-slaughter and stunning', *1st International Virtual Conference on Pork Quality*, Brazil, pp. 145–58.

Christensen, L. and Barton-Gade, P. (1997), 'New Danish developments in pig handling at abattoirs', *Fleischwirt.*, 77, 604–7.

Conte, S., Faucitano, L., Bergeron, R., Torrey, S., Gonyou, H. W., et al. (2015), 'Effects of season, truck type and location within truck on gastrointestinal tract temperature of market-weight pigs during transport', *J. Anim. Sci.*, 93, 5840–8.

Correa, J. A. (2011), 'Effect of farm handling and transport on physiological response, losses and meat quality of commercial pigs', *Adv. Pork Prod.*, 22, 249–56.

Correa, J. A., Torrey, S., Devillers, N., Laforest, J. P., Gonyou, H. W. and Faucitano, L. (2010), 'Effects of different moving devices at loading on stress response and meat quality in pigs', *J. Anim. Sci.*, 88, 4086–93.

Correa, J. A., Gonyou, H. W., Torrey, S., Widowski, T., Bergeron, R., et al. (2013), 'Welfare and carcass and meat quality of pigs being transported for 2 hours using two vehicle types during two seasons of the year', *Can. J. Anim. Sci.*, 93, 43–55

Correa, J. A., Gonyou, H. W., Torrey, S., Widowski, T., Bergeron, R., et al. (2014), 'Welfare of pigs being transported for long distance using a pot-belly trailer during winter and summer', *Animals*, 4, 200–13.

Council Regulation (EC) 1099/2009 (2009), 'Council Regulation No 1099/2009 on the protection of animals at the time of killing', *OJEU*, L303, 1–30.

CPC (2011), 'Animal care assessment', Canadian Pork Council, Ottawa, Canada. http://www.cqa-aqc.com/aca/documents/ACA-Animal-Care-Assessment.pdf.

Croft, P. G. (1952), 'Problem of electrical stunning', *Vet. Rec.*, 64, 255–8.

Dalla Costa, O. A., Faucitano, L., Coldebella, A., Ludke, J. V., Peloso, J. V., et al. (2007), 'Effects of the season of the year, truck type and location on truck on skin bruises and meat quality in pigs', *Livest. Sci.*, 107, 29–36.

Dalla Costa, F. A., Devillers, N., Paranhos da Costa, M. J. R. and Faucitano, L. (2016), 'Effects of applying preslaughter feed withdrawal at the abattoir on behaviour, blood parameters and meat quality in pigs', *Meat Sci.*, 119, 89–94.

Dalla Costa, O. A., Dalla Costa, F. A., Feddern, V., Dos Santos Lopes, L., et al. (2019), 'Risk factors associated with pig pre-slaughtering losses', *Meat Sci.*, 155, 61–8.

Dalmau, A. (2017), 'Development of a certification schema on animal welfare based on Welfare Quality protocols', *Newsl. Welf. Qual. Netw.*, 5, 2–3.

Dalmau, A., Temple, D., Rodríguez, P., Llonch, P. and Velarde, A. (2009), 'Application of the Welfare Quality® protocol at pig slaughterhouses', *Anim. Welf.*, 18, 497–505.

Dalmau, A., Llonch, P., Rodríguez, P., Ruíz-de-la-Torre, J. L., et al. (2010), 'Stunning pigs with different gas mixtures. Part 1: gas stability', *Anim. Welf.*, 19, 315-23.

Dewey, C., Haley, C., Widowski, T., Poljak, Z. and Friendship, R. (2009), 'Factors associated with in-transit losses of fattening pigs', *Anim. Welf.*, 18, 355-61.

Dokmanović, M., Velarde, A., Tomović,V., Glamočlija, N., Marković, R., et al. (2014), 'The effects of lairage time and handling procedure prior to slaughter on stress and meat quality parameters in pigs', *Meat Sci.*, 98, 220-6.

Edwards, L. N., Grandin, T. A., Engle, T. E., Porter, S. P., Ritter, M. J., et al. (2010), 'Use of exsanguination blood lactate to assess the quality of preslaughter pig handling', *Meat Sci.*, 86, 384-90.

Edwards, L. N, Engle, T. E., Grandin, T., Ritter, M. J., Sosnicki, A., et al. (2011), 'The effects of distance traveled during loading, lairage time prior to slaughter, and distance traveled to the stunning area on blood lactate concentration of pigs in a commercial packing plant', *Prof. Anim. Sci.*, 27, 485-91.

EFSA (2004), 'Welfare aspects of animal stunning and killing methods', Scientific Report of the Scientific Panel for Animal Health and Welfare on request from the Commission related to welfare aspects of animal stunning and killing methods (Question No. EFSA-Q-2003-093). European Food Safety Authority (AHAW 04-027).

EFSA (2013), 'Scientific Opinion on monitoring procedures at slaughterhouses for pigs', *EFSA J.*, 11(12), 3523.

Eyes on Animals (2019), 'Improving animal welfare in pig slaughterhouses'. http://www .eyesonanimals.com/wp-content/uploads/2016/06/Animal-welfare-in-pig-slaught erhouses-how-to-reduce-stress-suffering-and-ease-handling-aanp-1.pdf (accessed on 14 May 2019).

Faucitano, L. (2001), 'Causes of skin damage to pig carcasses', *Can. J. Anim. Sci.*, 81, 39-45.

Faucitano, L. (2010), 'Effects of lairage and slaughter conditions on animal welfare and pork quality', *Can. J. Anim. Sci.*, 90, 461-9.

Faucitano, L. (2018), 'Preslaughter handling practices and their effects on animal welfare and pork quality', *J. Anim. Sci.*, 96, 728-38.

Faucitano, L. and Goumon, S. (2018), 'Transport to slaughter and associated handling', *in* M. Špinka (Ed.), *Advances in Pig Welfare*, London, UK: Woodhead Publishing, pp. 261-93.

Faucitano, L. and Lambooij, E. (2019), 'Transport of pigs', *in* T. Grandin (Ed.), *Livestock Transport and Handling* (5th edn.), Wallingford, UK: CABI Publishing, pp. 302-27.

Fitzgerald, R. F., Stalder, K. J., Matthews, J. O., Schultz-Kaster, C. M. and Johnson, A. K. (2009), 'Factors associated with fatigued, injured, and dead pig frequency during transport and lairage at a commercial abattoir', *J. Anim. Sci.*, 87, 1156-66.

Fox, J., Widowski, T., Torrey, S., Nannoni, E., Bergeron, R., et al. (2014), 'Water sprinkling market pigs in a stationary trailer. 1. Effects on pig behaviour, gastrointestinal tract temperature and trailer micro-climate', *Livest. Sci.*, 160, 113-23.

Fraqueza, M. J., Roseiro, L. C., Almeida, J., Matias, E., Santos, C., et al. (1998), 'Effects of lairage temperature and holding time on pig behavior and on carcass and meat quality', *Appl. Anim. Behav. Sci.*, 60, 317-30.

Gallo, C., Faucitano, L. and Gerritzen, M. (2016), 'Effects of preslaughter handling on carcass and meat quality', in M. Raj and A. Velarde (Eds), *Animal Welfare at Slaughter: a Practitioner Guide*, Sheffield, UK: 5m Publishing, pp. 251-69.

Garcia, A. and McGlone, J. J. (2015), 'Loading and unloading finishing pigs: effects of bedding types, ramp angle, and bedding moisture', *Animals*, 5, 13-26.

Gerritzen, M. A., Hindle, V. A., Steinkamp, K. and Reimert, H. G. M. (2013), 'The effect of reduced loading density on pig welfare during long distance transport', *Anim.* 7, 1849-57.

Geverink, N. A., Engel, B., Lambooij, E. and Wiegant, V. M. (1996), 'Observations on behavior and skin damage of slaughter pigs and treatment during lairage', *Appl. Anim. Behav. Sci.*, 50, 1-13.

Geverink, N. A., Buhnemann, A., Van de Burgwal, J. A., Lambooij, E., Blokhuis, H. J., et al. (1998), 'Responses of slaughter pigs to transport and lairage sounds', *Physiol. Behav.*, 63, 667-73.

Gonyou, H. W. and Brown, J. (2012), 'Reducing stress and improving recovery from handling during loading and transport of market pigs', Final Report submitted to Alberta Livestock and Meat Agency, Edmonton, Canada, 40p.

Goumon, S. and Faucitano, L. (2017), 'Influence of loading handling and facilities on the subsequent response to pre-slaughter stress in pigs', *Livest. Sci.*, 200, 6-13.

Goumon, S., Brown, J. A., Faucitano, L., Bergeron, R., Widowski, T. M., et al. (2013a), 'Effects of transport duration on maintenance behavior, heart rate and gastrointestinal tract temperature of market-weight pigs in 2 seasons', *J. Anim. Sci.*, 91, 4925-35.

Goumon, S., Faucitano, L., Bergeron, R., Crowe, T., Connor, M. L., et al. (2013b), 'Effect of ramp configuration on easiness of handling, heart rate and behavior of near-market pigs at unloading', *J. Anim. Sci.*, 91, 3889-98.

Grandin, T. (1990), 'Design of loading facilities and holding pens', *Appl. Anim. Behav. Sci.*, 28, 187-201.

Grandin, T. (2010), 'The importance of measurement to improve the welfare of livestock, poultry and fish', in T. Grandin (Ed.), *Improving Animal Welfare: a Practical Approach*, Wallingford, UK: CABI Publishing, pp. 1-20.

Grandin, T. (2012), *Recommended Animal Handling Guidelines and Audit Guide: a Systematic Approach to Animal Welfare*, Washington, DC: American Meat Institute Foundation, 108p.

Grandin, T. (2016a), 'Transport fitness of cull sows and boars: a comparison of different guidelines of fitness for transport', *Animals*, 6, 77 (open access). doi: 10.3390/ani8070124.

Grandin, T. (2016b), 'Practical methods to improve animal handling and restraint', *in* A. Velarde and M. Raj (Eds), *Animal Welfare at Slaughter*, Sheffield, UK: 5M Publishing, pp. 71-90.

Grandin, T. (2017a), 'How to work with large meat buyers to improve animal welfare', *in* P. P. Purslow (Ed.), *New Aspects of Meat Quality: from Genes to Ethics*, London, UK: Woodhead Publishing, pp. 569-79.

Grandin, T. (2017b), *Recommended Animal Handling Guidelines and Audit Guide: Systematic Approach to Animal Welfare*, Washington DC: North American Meat Institute Foundation, 108p.

Grandin, T. (2018), 'Welfare problems in cattle, pigs, and sheep that persist even though scientific research clearly shows how to prevent them', *Animals*, 8, 124 (open access). doi: 10.3390/ani8070124.

Grandin, T. and Smith, G. C. (2004), 'Animal welfare and humane slaughter'. http://www.grandin.com/references/humane.slaughter.html (accessed on 23 September 2019).

Griot, B., Boulard, J., Chevillon, P. and Kerisit, R. (2000), 'Des restrainers à bande pour le bienêtre et la qualité de la viande', *Viandes et Produits Carnés*, 3, 91-7.

Guàrdia, M. D., Gispert, M. and Diestre, A. (1996), 'Mortality rates during transport and lairage in pigs for slaughter', *Meat Focus Inter.*, 10, 362-6.

Guàrdia, M. D., Estany, J., Balasch, S., Oliver, M. A., Gispert, M., et al. (2004), 'Risk assessment of PSE condition due to pre-slaughter conditions and RYR1 gene in pigs', *Meat Sci.*, 67, 471-8.

Guàrdia, M. D, Estany, J., Balasch, S., Oliver, M. A., Gispert, M., et al. (2005), 'Risk assessment of DFD meat due to pre-slaughter conditions in pigs', *Meat Sci.*, 70, 709-16.

Guàrdia, M. D., Estany, J., Balasch, S., Oliver, M. A., Gispert, M., et al. (2009), 'Risk assessment of skin damage due to pre-slaughter conditions and RYR1 gene in pigs', *Meat Sci.*, 81, 745-51.

Haley, C., Dewey, C. E., Widowski, T. and Friendship, R. (2008), 'Association between in-transit losses, internal trailer temperature, and distance travelled by Ontario market hogs', *Can. J. Vet. Res.*, 72, 385-9.

Haley, C., Dewey, C. E., Widowski, T. and Friendship, R. (2010), 'Relationship between estimated finishing-pig space allowance and in transit loss in a retrospective survey for 3 packing plant in Ontario in 2003', *Can. J. Vet. Res.*, 74, 178-84.

Hambrecht, E., Eissen, J. J., Newman, D. J., Smits, C. H., Verstegen, M. W., et al. (2005), 'Preslaughter handling effects on pork quality and glycolytic potential in two muscles differing in fiber type composition', *J. Anim. Sci.*, 83, 900-7.

Hoenderken, R. (1978), 'Electrical stunning of pigs for slaughter', Doctoral thesis, University of Utrecht, NL.

Holst, S. (2001), 'Carbon dioxide stunning of pigs for slaughter – practical guidelines for good animal welfare', *47th International Congress of Meat Science and Technology*, Krakow, Poland, pp. 48-54.

Honkavaara, M. (1989), 'Influence of lairage on blood composition of pig and on the development of PSE pork', *J. Agric. Sci. Finland*, 61, 433-40.

Humane Farm Animal Care (2018), 'Humane farm animal care standards', Humane Farm Animal Care, Middleburg, VA, 31p. www.certifiedhumane.org.

Huynh, T. T. T., Aarnink, A. J. A., Truong, C. T., Kemp, B. and Verstegen, M. W. A. (2006), 'Effects of tropical climate and water cooling on growing pigs' responses', *Livest. Sci.*, 104, 278-91.

Kephart, K. B., Harper, M. T. and Raines, C. R. (2010), 'Observations of market pigs following transport to a packing plant', *J. Anim. Sci.*, 88, 2199-203.

Knowles, T. G., Brown, S. N., Edwards, J. E. and Warriss, P. D. (1998), 'Ambient temperature below which pigs should not be continuously showered in lairage', *Vet. Rec.*, 143, 575-8.

Koketsu Y. (2000), 'Factors associated with increased sow mortality in North America', *Proceedings of the American Association of Swine Practitioners*, pp. 419-20.

Lambooij, E. (1988), 'Road transport of pigs over a long distance: some aspects of behaviour, temperature and humidity during transport and some effects of the last two factors', *Anim. Prod.*, 46, 257-63.

Lambooij, E. (2014), 'Transport of pigs', in T. Grandin (Ed.), *Livestock Handling and Transport*, Wallingford, UK: CABI Publishing, pp. 280-97.

Lambooij, E. and Engel, B. (1991), 'Transport of slaughter pigs by road over a long distance: some aspects of loading density and ventilation', *Livest. Prod. Sci.*, 28, 163-74.

Lambooij, E., Garssen, G. J., Walstra, P., Mateman, F. and Merkus, G. S. M. (1985), 'Transport of pigs by car for two days: some aspects of watering and loading density', *Livest. Prod. Sci.*, 13, 289-99.

Leheska, J. M., Wulf, D. M. and Maddock, R. J. (2003), 'Effects of fasting and transportation on pork quality development and extent of postmortem metabolism', *J. Anim. Sci.*, 80, 3194-202.

Long, V. P. and Tarrant, P. V. (1990), 'The effect of pre-slaughter showering and post-slaughter rapid chilling on meat quality in intact pork sides', *Meat Sci.*, 27, 181-95.

Malena, M., Voslárová, E., Kozák, A., Belobrádek, P., Bedánová, I., et al. (2007), 'Comparison of mortality rates in different categories of pigs and cattle during transport for slaughter', *Acta Vet. Brno*, 76, 109-16.

Martin, J. E., Christensen, K., Vizzier-Thaxton, Y. and McKeegan, D. E. F. (2016), 'Effects of analgesic intervention on behavioural responses to Low Atmospheric Pressure Stunning', *Appl. Anim. Behav. Sci.*, 180, 157-65.

McGee, M., Johnson, A. K., O'Connor, A. M., Tapper, K. R. and Millman, S. T. (2016), 'An assessment of swine marketed through buying stations and development of fitness for transport guidelines', *J. Anim. Sci.*, 94, 9.

McGlone, J. J., Johnson, A. K., Sapkota, A. and Kephart, R. K. (2014a), 'Establishing trailer ventilation (boarding) requirements for finishing pigs during transport', *Animals*, 4, 515-23.

McGlone, J. J., Johnson, A. K., Sapkota, A. and Kephart, R. K. (2014b), 'Transport of market pigs: improvements in welfare and economics', *in* T. Grandin (Ed.), *Livestock Handling and Transport*, Wallingford, UK: CABI Publishing, pp. 298-314.

Moss, B. W. (1978), 'Some observations on the activity and aggressive behavior of pigs when penned prior to slaughter', *Appl. Anim. Ethol.*, 4, 323-39.

Mota-Rojas, D., Becerril, M., Lemus, C., Sánchez, P., González, M., Olmos, S. A., et al. (2006), 'Effects of mid-summer transport duration on pre- and post-slaughter performance and pork quality in Mexico', *Meat Sci.*, 73, 404-12.

Nanni Costa, L., Lo Fiego, D. P., Dall'Olio, S., Davoli, R. and Russo, V. (2002), 'Combined effects of pre-slaughter treatments and lairage time on carcass and meat quality in pigs of different halothane genotype', *Meat Sci.*, 61, 41-7.

Nannoni, E., Widowski, T. M., Torrey, S., Fox, J., Rocha, L. M., et al. (2014), 'Water sprinkling market pigs in a stationary trailer. 2. Effects on selected exsanguination blood parameters and carcass and meat quality variation', *Livest. Sci.*, 160, 124-31.

Nannoni, E., Liuzzo, G., Serraino, A., Giacometti, F., Martelli, G., et al. (2016), 'Evaluation of pre-slaughter losses of Italian heavy pigs', *Anim. Prod. Sci.* (open-access). doi: 10.1071/AN15893.

Nowak, B., Mueffling, T. V. and Hartung, J. (2007), 'Effect of different carbon dioxide concentrations and exposure times in stunning of slaughter pigs: impact on animal welfare and meat quality', *Meat Sci.* 75, 290-8.

Pereira, T., Titto, E. A., Conte, S., Devillers, N., Sommavilla, R., et al. (2018), 'Use of fan-misters bank for cooling pigs kept in a stationary trailer before unloading: effects on trailer microclimate, and pig behavior and physiological response', *Livest. Sci.*, 216, 67-74.

Peterson, E., Remmenga, M., Hagerman, A. D. and Akkina, J. E. (2017), 'Use of temperature, humidity, and slaughter condemnation data to predict increases in transport losses in three classes of swine and resulting foregone revenue', *Front. Vet. Sci.*, 4, 67 (open access). doi: 10.3389/fvets.2017.00067.

Pilcher, C. M., Ellis, M., Rojo-Gomez, A., Curtis, S. E., Wolter, B. F., et al. (2011), 'Effects of floor space during transport and journey time on indicators of stress and transport losses in market weight pigs', *J. Anim. Sci.*, 89, 3809-18.

Rabaste, C., Faucitano, L., Saucier, L., Foury, D., Mormède, P., et al. (2007), 'The effects of handling and group size on welfare of pigs in lairage and its influence on stomach weight, carcass microbial contamination and meat quality variation', *Can. J. Anim. Sci.*, 87, 3-12.

Raj, A. B. M. and Gregory, N. G. (1996), 'Welfare implications of the gas stunning of pigs. Stress of induction of anaesthesia', *Anim. Welf.*, 5, 71-8.

Raj, A. B. M. and Velarde, A. (2016), *Animal Welfare at Slaughter: A Practitioner Guide*, Sheffield, UK: 5m Publishing.

Raj, A. B. M., Johnson, S. P., Wotton, S. B. and McInstry, J. L. (1997), 'Welfare implications of gas stunning of pigs 3. Time to loss of somatosensory evoked potentials and spontaneous electrocorticogram of pigs during exposure to gases', *Br. Vet. J.*, 153, 329-40.

Renaudeau, D., Gourdine, J. and St-Pierre, N. (2011), 'A meta-analysis of the effects of high ambient temperature on growth performance of growing-finishing pigs', *J. Anim. Sci.*, 89, 2220-30.

Rioja-Lang, F. C., Brown, J. A., Brockhoff, E. J. and Faucitano, L. (2019), 'A review of swine transportation research on priority welfare issues: a Canadian perspective', *Front. Vet. Sci.*, 6, 36 (open access). doi: 10.3389/fvets.2019.00036.

Ritter, M. J., Ellis, M., Brinkmann, J., DeDecker, J. M., Keffaber, K. K., et al. (2006), 'Effect of floor space during transport of market-weight pigs on the incidence of transport losses at the packing plant and the relationships between transport conditions and losses', *J. Anim. Sci.*, 84, 2856-64.

Ritter, M. J., Ellis, M., Bowman, R., Brinkmann, J., Curtis, S. E., et al. (2008), 'Effects of season and distance moved during loading on transport losses of market-weight pigs in two commercially available types of trailer', *J. Anim. Sci.*, 86, 3137-45.

Ritter, M., Rincker, P. and Carr, S. (2012), 'Pig handling and transportation strategies utilized under U.S. commercial conditions', London Swine Conference, London, Canada, pp. 109-20.

Rocha, L. M., Bridi, A. M., Foury, A., Mormède, P., Weschenfelder, A. V., et al. (2013), 'Effects of ractopamine administration and castration method on the response to pre-slaughter stress and carcass and meat quality in pigs of two Piétrain genotypes', *J. Anim. Sci.*, 91, 3965-77.

Rocha, L. M., Dionne, A., Saucier, L., Nannoni, E. and Faucitano, L. (2015), 'Hand-held lactate analyzer as a tool for the real-time measurement of physical fatigue before slaughter and pork quality prediction', *Anim.* 9, 707-14.

Rocha, L. M., Velarde, A., Dalmau, A., Saucier, L. and Faucitano, L. (2016), 'Can the monitoring of animal welfare parameters predict pork meat quality variation through the supply chain (from farm to slaughter)?', *J. Anim. Sci.*, 94, 359-76.

Rodríguez, P., Dalmau, A., Ruiz-de-la-Torre, J. L., Manteca, X., Jensen, E. W., et al. (2008), 'Assessment of unconsciousness during carbon dioxide stunning in pigs', *Anim. Welf.*, 17, 341-9.

Rydhmer, L., Zamaratskaia, G., Andersson, H. K., Algers, B., Guillemet, R., et al. (2006), 'Aggressive and sexual behavior of growing and finishing pigs reared in groups, without castration', *Acta Agric. Scand., Sect. A Anim. Sci.*, 56, 109-19.

Santos, C., Almeida, J. M., Matias, E. C., Fraqueza, M. J., et al. (1997), 'Influence of lairage environmental conditions and resting time on meat quality in pigs', *Meat Sci.*, 45, 253-62

SCAHAW (Scientific Committee on Animal Health and Animal Welfare) (2011), 'Scientific opinion concerning the welfare of animals during transport', *EFSA J.*, 9, 1966.

Scheeren, M. B., Gonyou, H. W., Brown, J., Weschenfelder, A. V. and Faucitano, L. (2014), 'Effects of transport time and location within truck on skin bruises and meat quality of market weight pigs in two seasons', *Can. J. Anim. Sci.*, 94, 71–8.

Schwartzkopf-Genswein, K. S., Faucitano, L., Dadgar, S., Shand, P., Gonzàlez, L. A., et al. (2012), 'Road transportation of cattle, swine and poultry in North America and its impact on animal welfare, carcass and meat quality: a review', *Meat Sci.*, 92, 227–43.

Shen, Q. W., Means, W. J., Thompson, S. A., Underwood, K. R., Zhu, M. J., et al. (2006), 'Pre-slaughter transport, AMP-activated protein kinase, glycolysis, and quality of pork loin', *Meat Sci.*, 74, 388–95.

Sommavilla, R., Faucitano, L., Gonyou, H. W., Seddon, Y., Bergeron, R., et al. (2017), 'Season, transport duration and trailer compartment effects on blood stress indicators in pigs: relationship to environmental, behavioural and other physiological factors, and pork quality traits', *Animals*, 7, 8 (open access). doi: 10.3390/ani7020008.

Sutherland, M. A., McDonald, A. and McGlone, J. J. (2009), 'Effects of variations in the environment, length of journey and type of trailer on the mortality and morbidity of pigs being transported to slaughter', *Vet. Rec.*, 165, 13-18.

Talling, J. C., Waran, N. K., Wathes, C. M. and Lines, J. A. (1996), 'Behavioural and physiological responses of pigs to sound', *Appl. Anim. Behav. Sci.*, 48, 187-202.

Tanida, H., Miura, A., Tanaka, T. and Yoshimoto, T. (1996), 'Behavioral responses of pig to darkness and shadows', *Appl. Anim. Behav. Sci.*, 49, 173-83.

Thodberg, K., Fogsgaard, K. K. and Herskin, M. S. (2019), 'Transportation of cull sows – deterioration of clinical condition from departure and until arrival at the slaughter plant', *Front. Vet. Sci.*, 6, 28 (open access). doi: 10.3389/fvets.2019.00028.

Torrey, S., Bergeron, R., Gonyou, H. W., Widowski, T. M., Lewis, N., et al. (2013a), 'Transportation of market-weight pigs: 1. Effect of season and truck type on behavior with a 2-hour transport', *J. Anim. Sci.*, 91, 2863-71.

Torrey, S., Bergeron, R., Faucitano, L., Widowski, T. M., Lewis, N., et al. (2013b), 'Transportation of market-weight pigs: 2. Effect of season and animal location in the truck on behavior with an 8-hour transport', *J. Anim. Sci.*, 91, 2872-8.

TQA (2016), '*National Pork Board Transport Quality Assurance Handbook*', Version 6. http://www.pork.org/tqa-certification/tqa-program-materials/ (accessed on 13 May 2019).

Van de Perre, V., Permentier, L., de Bie, S., Verbeke, G. and Geers, R. (2010), 'Effect of unloading, lairage, pig handling, stunning and season on pH of pork', *Meat Sci.*, 86, 931-7.

Van der Wal, P. G., Engel, B. and Reimert, H. G. M. (1999), 'The effect of stress, applied immediately before stunning, on pork quality', *Meat Sci.*, 53, 101-6.

Van Putten, G. and Elshof, W. J. (1978), 'Observations on the effect of transportation on the well-being and lean quality of slaughter pigs', *Anim. Reg. Stud.*, 1, 247-71.

Vecerek, V., Malena, M., Malena Jr., M., Voslarova, E. and Chloupek, P. (2006), 'The impact of the transport distance and season on losses of fattened pigs during transport to the slaughterhouse in the Czech Republic in the period from 1997 to 2004', *Vet. Med.*, 51, 21-8.

Velarde, A. and Dalmau, A. (2012), 'Animal welfare assessment at slaughter in Europe: moving from inputs to outputs', *Meat Sci.*, 92, 244-51.

Velarde, A., Gispert, M., Faucitano, L., Manteca, X. and Diestre, A. (2000), 'Survey of the effectiveness of stunning procedures used in Spanish pig abattoirs', *Vet. Rec.*, 146, 65-8.

Velarde, A., Gispert, M., Faucitano, L., Alonso, P., Manteca, X., et al. (2001), 'Effects of the stunning procedure and the halothane genotype on meat quality and incidence of haemorrhages in pigs', *Meat Sci.* 58, 313-19.

Velarde, A., Cruz, J., Gispert, M., Carrión, D., Ruiz-de-la-Torre, J. L., et al. (2007), 'Aversion to carbon dioxide stunning in pigs: effect of the carbon dioxide concentration and the halothane genotype', *Anim. Welf.*, 16, 513-22.

Vermeulen, L., Van de Perre, V., Permentier, L., De Bie, S., Verbeke, G., et al. (2015), 'Sound levels above 85 dB pre-slaughter influence pork quality', *Meat Sci.*, 100, 269-74.

Vitali, A., Lana, E., Amadori, M., Bernabucci, U., Nardone, A., et al. (2014), 'Analysis of factors associated with mortality of heavy slaughter pigs during transport and lairage', *J. Anim. Sci.*, 92, 5134-41.

Warriss, P. D. (1996), 'The consequences of fighting between mixed groups of unfamiliar pigs before slaughter', *Meat Focus Int.*, 5, 89-92.

Warriss, P. D. (2003), 'Optimal lairage times and conditions for slaughter pigs: a review', *Vet. Rec.*, 153, 170-6.

Warriss, P. D. and Brown, S. N. (1985), 'The physiological responses of fighting between in pigs and the consequences for meat quality', *J. Sci. Food Agric.*, 36, 87-92.

Warriss, P. D. and Brown, S. N. (1994), 'A survey of mortality in slaughter pigs during transport and lairage', *Vet. Rec.*, 134, 513-15.

Warriss, P. D. and Wilkins, L. J. (1987), 'Exsanguination in meat animals', *Seminar Preslaughter Stunning of Food Animals*, Brussels, Belgium, pp. 150-8.

Warriss, P. D., Brown, S. N., Bevis, E. A. and Kestin, S. C. (1990), 'The influence of pre-slaughter transport and lairage on meat quality in pigs of two genotypes', *Anim. Prod.*, 50, 165-72.

Warriss, P. D., Bevis, E. A., Edwards, J. E., Brown, S. N. and Knowles, T. G. (1991), 'Effect of the angle of slope on the ease with which pigs negotiate loading ramps', *Vet. Rec.*, 128, 419-21.

Warriss, P. D., Brown, S. N., Adams, S. J. M. and Corlett, I. K. (1994), 'Relationships between subjective and objective assessments of stress at slaughter and meat quality in pigs', *Meat Sci.*, 38, 329-40.

Weeks, C. A. (2008), 'A review of welfare in cattle, sheep, and pig lairages, with emphasis on stocking rates, ventilation and noise', *Anim. Welf.*, 17, 275-84.

Weeks, C. A., Brown, S. N., Warriss, P. D., Lane, S., Heasman, L., et al. (2009), 'Noise levels in lairages for cattle, sheep and pigs in abattoirs in England and Wales', *Vet. Rec.*, 165, 308-14.

Werner, C., Reiners, K. and Wicke, M. (2007), 'Short as well as long transport duration can affect the welfare of slaughter pigs', *Anim. Welf.*, 16, 385-9.

Weschenfelder, A. V., Torrey, S., Devillers, N., Crowe, T., Bassols, A., et al. (2012), 'Effects of trailer design on animal welfare parameters and carcass and meat quality of three Pietrain crosses being transported over a long distance', *J. Anim. Sci.*, 90, 3220-4676.

Weschenfelder, A. V., Torrey, S., Devillers, N., Crowe, T., Bassols, A., et al. (2013), 'Effects of trailer design on animal welfare parameters and carcass and meat quality of three Pietrain crosses being transported over a short distance', *Livest. Sci.*, 157, 234-44.

Wesoly, R., Jungbluth, I., Stefanski, V. and Weiler, U. (2015), 'Pre-slaughter conditions influence skatole and androstenone in adipose tissue of boars', *Meat Sci.*, 99, 6067.

WQ* (2009), 'Welfare Quality assessment protocol for pigs', Welfare Quality Consortium, Lelystad, the Netherlands, 122p.

Wotton, S. B. and Gregory, N. G. (1986), 'Pig slaughtering procedures: time to loss of brain responsiveness after exsanguination or cardiac arrest', *Res. Vet. Sci.*, 40, 148-51.

Xiong, Y, Green, A. and Gates, R. S. (2015), 'Characteristics of trailer thermal environment during commercial swine transport managed under U.S. industry guidelines', *Animals*, 5, 226-44.

Zhao, Y., Lu, X., Mo, D., Chen, Q. and Chen, Y. (2015), 'Analysis of reasons for sow culling and seasonal effects on reproductive disorders in southern China', *Anim. Rep. Sci.*, 159, 191-7.

Zurbrigg, K., van Dreumel, T., Rothschild, M. F., Alves, D., Friendship, R., et al. (2017), 'Pig-level risk factors for in-transit losses in swine: a review', *Can. J. Anim. Sci.*, 97, 339-46.

Chapter 4

Improving welfare in poultry slaughter

Dorothy McKeegan, Institute of Biodiversity, Animal Health and Comparative Medicine, University of Glasgow, UK; and Jessica Martin, The Royal (Dick) School of Veterinary Studies and The Roslin Institute, University of Edinburgh, UK

1 Introduction

In 2018, 68.7 billion chickens were slaughtered for food globally. Based on current trends, this number is predicted to increase to 72.4 billion by 2020, based on an average yearly 2.6% increase (Food and Agriculture Organization of the United Nations (FAO), 2020). In 2018, in the United Kingdom, 1.14 billion broilers were slaughtered across 50 operating slaughter houses and 57 million spent hens were processed in 9 plants (Defra, 2019; Food and Agriculture Organization of the United Nations (FAO), 2020). In comparison for the same year, just over 9 billion broilers were slaughtered in the United States, across approximately 320 plants (United States Department of Agriculture, 2019). Excluding farmed and wild-caught fish, poultry production represents, by a large margin, the greatest number of individual animals killed by humans for food production. This reflects their status as the most numerous terrestrial production animals. Because of the number of sentient animals involved, the manner of their deaths is extremely important, since it may be argued that our ethical responsibilities to animals under our care extend to providing them with good lives and good deaths (Singer, 2016).

For clarity, it is helpful to explain the commonly used terminology in relation to the slaughter of chickens. Slaughter may be defined as the killing of animals intended for human consumption, usually involving exsanguination (bleeding out). During slaughter, animals are usually stunned prior to exsanguination to render them unconscious (though certain types of religious slaughter do

http://dx.doi.org/10.19103/AS.2020.0078.14

not apply stunning, and we will discuss the welfare implications and scale of religious slaughter later in this chapter). Stunning refers to any intentionally induced process which causes loss of consciousness and sensibility with the minimum associated stress and pain, including any process resulting in instantaneous death (European Commission, 2017; Nielsen et al., 2019a). Stunning improves welfare at slaughter because its aim is to prevent the animal from experiencing pain from the neck cut and its own death via blood loss and associated brain hypoxia. Stunning methods can result in a recoverable state (if exsanguination does not occur), referred to as a simple stun or can induce a 'non recovery state', such that the animal would not recover, even in the absence of exsanguination.

The term 'killing' refers to any intentionally induced process which causes the death of an animal (EC, 2009). Emergency killing means the killing of animals which are injured or have a disease associated with severe pain or suffering and where there is no other practical possibility to alleviate this pain or suffering (EC, 2009). Killing methods do not have to include a pre-kill stun step. Therefore the animals may be conscious (if not already unconscious as are result of sickness or injury) when the killing technique is applied and will potentially be able to experience negative emotions (e.g. pain and stress), if the killing method does not render them immediately unconscious. The term 'culling' is used technically to describe identifying and removing individuals from a group (Fetrow et al., 2006), but it has become common to use this term also to describe the removal and killing (usually limited to non-slaughter purposes) of individuals. Euthanasia is defined literally as providing a 'good death' and/or providing a death that is in the animal's interest (usually to relieve suffering). In practice this usually means killing an animal as painlessly as possible with an overdose of anaesthetic. Strictly, the term is only applicable to sick or injured animals (veterinary patients) or animals used in scientific research (Leary et al., 2020, 2007). Euthanasia is often misused as a term for describing killing of animals in other contexts. Loss of consciousness describes a transition in which the animal moves from a state of conscious awareness (connectedness to the environment and responsiveness to stimuli, for example, painful events) to unconsciousness which occurs when the ability to maintain an awareness of self and environment is lost (involving a complete or near-complete lack of responsiveness to environmental stimuli). The existence of an inducible state of unconsciousness and associated changes in brain state and resulting electroencephalogram (EEG) characteristics appear to be common across vertebrates (Pierre et al., 2018; Verhoeven et al., 2014).

In many jurisdictions, animals can only be exsanguinated after stunning, and in accordance with certain allowed methods, with some exceptions authorised for religious rites. For effective stunning, the induced loss of consciousness and sensibility must be maintained until the death of the animal.

In addition to slaughter, large numbers of farmed poultry are killed for other reasons, primarily on-farm due to individual ill health, non-competitive growth, emergency killing for disease control or depopulation (killing of low-value poultry at the end of their productive life) (Gerritzen et al., 2000; Martin et al., 2019b; McKeegan et al., 2011). These activities are associated with related but also distinct welfare challenges and are reviewed elsewhere (McKeegan, 2018; Nielsen et al., 2019b; Sparrey et al., 2014). In this chapter, we will focus on the welfare challenges presented by routine slaughter of chickens for food production. This includes chickens reared for meat (broilers) and also laying hens (so-called spent hens, which are usually slaughtered at the end of the egg production cycle), though it should be noted that the vast majority of available research relates to broilers. We will examine the various methods available and in current use, discussing their welfare costs and benefits. Given the large literature already available on some types of stunning and slaughter, and existing thorough reviews of the physiological basis for stunning methods (Berg and Raj, 2015; Blokhuis et al., 2004; Bøtner et al., 2012; Nielsen et al., 2019a; Raj, 1998; Shields and Raj, 2010), we will focus on outstanding welfare issues, possible opportunities to address these and examine the potential of emerging approaches. The slaughter procedure necessarily follows catching and transportation, processes which also have specific welfare impacts. Again, these are reviewed elsewhere (see Bøtner et al., 2011; Kettlewell and Mitchell, 1994; Knowles and Broom, 1990; Mitchell and Kettlewell, 1994) and will not be included here. We discuss welfare issues related to lairage (the holding period prior to slaughter, discussed briefly in Chapter 13 of this volume) and pre-slaughter handling, when directly relevant to the experience of birds. We will begin by briefly outlining some relevant regulatory frameworks, with a focus on the European Union which is widely recognised to have the most stringent legal protection for animals at the time of killing (EC, 2009). We will then discuss current and emerging methods, concluding with prospects for improvement of welfare based on available systems and identification of knowledge gaps for research.

1.1 Regulatory framework

Most countries have regulations that relate to animal welfare and/or meat hygiene at slaughter. Description of the full range of international regulation relevant to this issue is beyond the scope of this chapter, but we will briefly outline some relevant regulatory features, with a focus on the European Union (EU) which imposes detailed regulatory requirements on member states (EC, 2009). In the EU, Council Regulation (EC) No 1099/2009 on the protection of animals at the time of killing regulates the slaughter of animals. As a regulation it automatically became law in member states, with no transposition or scope for local interpretation. However, some member states retain national legislation which may provide additional protection at slaughter. Regulation 1099/2009 requires the use of approved stunning methods for poultry – these include electrical water bath stunning (with minimum electrical requirements) and controlled-atmosphere stunning (including a defined range of allowed gas mixtures and Low Atmospheric Pressure Stunning (LAPS) parameters). The regulation also imposes requirements for the layout, construction and equipment within slaughter houses, the appointment of an Animal Welfare Officer (accountable for implementing animal welfare measures) and certificates of competence for training of staff who handle live animals. Each member state has the responsibility for enforcement of legislation within its territory, but the DG for Health and Food Safety has an overall role for carrying out audits and inspections within member states. The requirement for allowable stunning methods extends to slaughter houses in third countries that export meat to the EU. The contents of Regulation 1099/2009 are evidence based, primarily supported by Opinions of the European Food Safety Authority (EFSA) on welfare at slaughter in 2004 and 2013 (Authie et al., 2013; Blokhuis et al., 2004; More et al., 2017) and the 'precautionary principle' – a strategy of caution when approaching issues of potential harm when extensive scientific knowledge on the matter is lacking.

Following EU exit, the UK will preserve the existing legislative framework initially via a general Withdrawal Bill, and it has been noted that EU exit could present opportunities to review and improve welfare standards (McCulloch, 2018). However, increased costs associated with more stringent welfare standards will need to be balanced with a desire to protect the competitiveness of UK producers in the EU and global markets (European Union, 2017). On exit from the EU, the UK will remain a member of the Council of Europe, an international organisation with 47 member states, which is distinct from the EU. The Council of Europe cannot make binding laws, but it has produced six conventions on animal welfare (including one on slaughter (European Council, 1979)) that act as a framework for animal welfare standards.

Although the United States is the world's largest poultry meat producer with 18% of global output (followed by China, Brazil and the Russian Federation), there are currently no federal regulations to control or safeguard the welfare of animals used in agriculture. An Animal Welfare Act (1966) is in place, but it applies only to animals kept for non-farming purposes. State laws regulate animal welfare in some parts of the country, but no such legislation applies to poultry in the three major poultry-producing states (Georgia, Alabama and Arkansas). Instead, voluntary welfare guidelines (those of the US National Chicken Council) are adopted in the absence of legislation, but these refer primarily to rearing conditions for chickens (e.g. stocking density limits), and not slaughter conditions (National Chicken Council (NCC), 2011). Poultry are not included in the Humane Slaughter Act (US Department of Agriculture, 1958) and electrical stunning (which is used almost universally in the United States) is not subject to stringent technical requirements (as in the EU). Oversight of slaughter is provided by a Poultry Health Veterinarian or other authorised personnel who are required to perform a routine inspection of slaughter procedures to ensure that slaughter is being undertaken according to legislative requirements and good commercial practices. Similarly, China (a key poultry meat producer and the biggest egg producer with 42% of global production), has no legislation with regard to poultry killing and slaughter. Instead, province-specific guidelines are in development. For example, in 2016, Shandong Province issued China's first technical manual for the more humane killing of chickens, which included new standard requirements such as birds being stunned by either gas or electrical methods prior to killing. Where laws are not present, guidelines may inform the basis of welfare standards and practical implementation. For example, the World Organisation for Animal Health (OIE) has developed in its Terrestrial Animal Health Code a chapter on the slaughter of animals (chapter 7.5) in order to ensure welfare during pre-slaughter and slaughter processes for the majority of livestock species (OIE (World Animal Health Organization), 2019).

2 Lairage

Lairage is a holding period at the processing plant prior to slaughter (Gregory, 2008; Warriss et al., 2005), lasting from arrival until animals enter the slaughter process. Lairage serves multiple functions – it ensures a continuous and smooth supply of animals to the slaughter line, it provides for ante-mortem inspection to take place for public health and welfare assessment purposes, and it allows birds to rest following transport. The lairage period is usually a maximum of a few hours (often a similar length to the transport journey time), but is occasionally longer (Caffrey et al., 2017; Gregory, 2008; Warriss et al., 2005). Grilli et al. (2015) reported lairage times for 233 different batches of broilers as ranging between 0.2 and 9.4 h with a mean of 4 h, while another study (Jacobs et al., 2017) based on six processing plants found lairage durations from 15 min to 9 h, with a mean of approximately 4.5 h. There is evidence that lengthy lairage durations (more than 6 h) lead to heightened stress and associated metabolic disorders due to prolonged feed and water withdrawal, with significant welfare consequences (Rodrigues et al., 2017; Vieira et al., 2011b).

During lairage, birds are held in their transport containers which are stacked, having been unloaded from transport vehicles, usually by forklift trucks. The condition of the birds at lairage is dependent on multiple factors including rearing conditions, catching practices and transport circumstances. Lairage, therefore, represents the culmination of a series of welfare challenges, with transport the most severe and best characterised of these. Transport imposes multiple stressors (e.g. acceleration, motion, vibration, withdrawal of food and water, social disruption and noise) as well as potentially extreme thermal microenvironments (Mitchell and Kettlewell, 1994, 1998) and its welfare impact may be exacerbated by injuries sustained at catching (Warriss et al., 1992). Spent laying hens endure longer transport distances than broilers (due to fewer processing sites) and are more prone to skeletal injury from handling than broilers (Knowles and Broom, 1990; Knowles and Wilkins, 1998). The most basic index of welfare in lairage is the number of birds dead on arrival (DOA). The number of birds DOA is of interest primarily economically and various studies have reported DOA rates and attempted to determine causative factors. Most studies report DOA rates up to 0.5% (e.g. Bayliss and Hinton, 1990; Ekstrand, 1998; Nijdam et al., 2004; Warriss et al., 1992) with occasionally higher rates, for example, 2% in broilers and 6.6% in spent hens (Petracci et al., 2006). The influences on rates of DOA are complex and include transport duration, waiting time, bird sex (and therefore weight, heavier birds are at greater risk), stocking density, ambient temperature, whether flocks had been previously been thinned, season of the year and feed withdrawal times exceeding 6 h (Cockram et al., 2019; Nijdam et al., 2004; Villarroel et al., 2018; Warriss et al., 1992). Although dead birds may be identified at ante-mortem inspection, this involves only a small sample of animals, so the true extent of DOA is only known at the point of manual shackling either before electrical stunning or after controlled atmosphere stunning (CAS). At least one known available slaughter system has dispensed with lairage and unloads modules directly on to the slaughter line from the vehicle (Meyn, 2020). There is no published research on this practice, and its welfare consequences presumably represent a balance between the negative impact of omitting a rest period versus the positive aspects of reducing feed withdrawal duration and exposure to other challenges in lairage.

Deaths detected at lairage and at subsequent handling most likely reflect mortality during transport, but it has been noted that holding in transport crates at lairage may also generate stressful thermal microclimates (Hunter et al., 1998; Quinn et al., 1998). Ideally, lairage facilities should be designed to protect bird welfare (Grilli et al., 2015), but many lairage areas are open to the outside and therefore reflect ambient conditions which can present thermal challenges. Additionally, limited ventilation in closely stacked containers may be worsened compared to transport by the lack of passive air movement in the absence of forward motion of a vehicle. The same point applies to delayed unloading from the transport vehicle, during which temperature can rise rapidly (Warriss et al., 1999, 2005). Given the practice of vertical stacking of transport containers, solid crate floors are preferred as they prevent birds from being exposed to the droppings of those above; however, perforated floors promote air movement. Forced ventilation and/or climate control is recommended and increasingly utilised in lairages to ensure comfortable thermal conditions, especially in warmer climates, and has been shown to reduce heat stress and mortality (Vieira et al., 2010, 2011a). It is also recommended that transport and lairage activities are reactive to weather conditions, avoiding the hottest part of the day and providing shade from the sun (Warriss et al., 2005). Some processing plants mist transport modules with water in lairage to promote evaporative heat loss, reducing heat stress (Jiang et al., 2015). Space between stacked transport crates is important and should be increased in hot conditions to promote ventilation. Conversely, cold stress in lairage can be reduced by providing shelter from wind and stacking crates closer together. Dimmed or blue lighting (which is perceived by birds as a lower light intensity) is routinely used in lairage to calm the birds and promote resting behaviour (Fig. 1). There is no published evidence of the effect of this practice in lairage, but evidence from rearing environments supports the notion that modified lighting may reduce stress following disturbance and reduces activity (Lewis and Morris, 2000; Mohamed et al., 2014; Prayitno et al., 1997a,b).

The transition from lairage to the slaughter line also presents welfare challenges. Manipulation of containers should avoid tilting, dropping or shaking them (this also applies to unloading from the transport vehicle), and container design is crucial to minimise injury to birds from body parts being caught between containers or in crate openings during stacking and unstacking

Figure 1 Poultry lairage facility showing stacked transport modules and the use of dimmed blue lighting. Image courtesy of Henny Reimert and Marien Gerritzen, Wageningen Livestock Research.

(Nielsen et al., 2019a). Various methods are used to remove birds from transport crates, and most relevant to welfare are those that take place before stunning. Manual removal of birds from crates (e.g. to allow shackling for electrical stunning, see below) has the potential to cause pain and fear, particularly if birds are handled roughly or held by inappropriate body parts leading to skin lesions, wing fractures and bruising (Jacobs et al., 2017; Sparrey and Kettlewell, 1994). Crate design has a major impact on the success of removal of birds and can contribute to injury if the opening is narrow, and modular systems with larger openings reduce this risk (Tinker et al., 2004). Kittelsen et al. (2015) reported that more wing fracture injuries occurred during pre-slaughter handling than at catching and transport (increasing from 0.8% in lairage to 2.9% after shackling, based on broilers undergoing electrical stunning). Some stunning systems involve tipping of the container such that birds are released and fall onto a flat sprung surface, for delivery onto a conveyer belt (Raj and Tserveni-Gousi, 2000). This practice is likely to cause fear and risks injury, especially if birds pile up, though overcrowding can be reduced with correct conveyor speeds.

It is clear that there are multiple opportunities to improve welfare at lairage including optimising transport conditions, use of controlled temperature lairage environments, minimising the duration of lairage (and thus the food and water withdrawal period), optimised container design and careful bird handling. Appropriate handling in particular is recognised as a key factor in protecting welfare prior to slaughter (OIE (World Animal Health Organization), 2019) and there is evidence that a positive management attitude and staff training is crucial to raise awareness (Nielsen et al., 2019a).

3 Stunning methods

In this section the major methods of stunning applied to chickens at slaughter will be discussed. Essentially, these are electrical methods and those that modify the atmosphere. In the UK and Europe, CAS (gas stunning) is most common, while electrical stunning is used in the rest of the world (Defra, 2019; Nielsen et al., 2019a). Mechanical methods are used only as a backup measure for commercial slaughter and will be discussed briefly. Different stunning and slaughter methods have strengths and weaknesses with regard to welfare outcomes, and the relevant risks and hazards for each method were recently reviewed by EFSA (Nielsen et al., 2019a). They noted that two categories of hazards are apparent during stunning – those leading to negative welfare during induction of unconsciousness, and those resulting from a delay or failure to achieve loss of consciousness (Nielsen et al., 2019a). This second category relates to the risk of birds being conscious and therefore exposed to the harms of further processing, since in all cases stunning is followed by neck cutting and exsanguination. Blood loss is the cause of death if the stun is reversible, or may occur after death with stun-kill methods. For chickens, neck cutting is usually done automatically and ideally involves the severing of both carotid arteries (Gregory and Wotton, 1986); however, occasionally only one or partial cuts are made. Whether this is a welfare issue depends on the type of stunning used – if reversible, stunning duration should be sufficient to allow death from blood loss without recovery of consciousness. Whatever method is used, the imperative must be to avoid the severely painful and fearful situations of ineffective stunning, leading to persistence of consciousness during neck cutting, or recovery of consciousness during bleeding.

3.1 Electrical stunning

Electrical stunning involves application of an electrical current to the brain, with the intention of inducing unconsciousness immediately prior to exsanguination. This approach is based on the principle of electrical stimulation causing immediate generalised epileptiform activity in the brain and associated unconsciousness and insensibility, protecting welfare during the neck cut and bleeding-out period. Epileptiform activity represents a temporary disruption of normal brain function, consisting of a tonic (e.g. whole body stiffening) and clonic (e.g. convulsions) phase, followed by a period of neuronal exhaustion and then recovery (Raj and Tserveni-Gousi, 2000). There is evidence that chickens do not always show typical epileptiform activity in the brain after electrical stunning (Gregory and Wotton, 1987), but application of adequate current is followed by a period of quiescent or supressed activity (as measured by the EEG, a global measure of brain activity), indicating effective stunning (Raj, 1998). In research settings, various criteria have been used to determine the effectiveness of electrical stunning, including monitoring of EEG activity, abolition of somatosensory evoked potentials in the brain and induction of cardiac arrest (Berg and Raj, 2015; Raj, 1998). Depending on whether cardiac arrest is caused, electrical stunning (electronarcosis) can be a reversible stunning method, so a killing method must be rapidly applied to abolish the risk of recovery of consciousness.

In poultry, electrical stunning is normally applied using a water bath stunner (Fig. 2), whereby the birds are hung upside down on a moving line with their feet held in grounded metal shackles. Shackling involves manually unloading birds from transport containers, inversion of the body by operators and insertion of the legs into parallel metal slots (termed shackles) on a moving conveyor, which transports the birds to the water bath. The birds' heads are then dipped into an electrically charged water bath, closing the circuit and causing current to flow through the head and body to the shackle. As such, birds are potentially stunned as soon as their heads enter the water (Bilgili, 1999; Devos et al., 2018; Prinz et al., 2010a), which allows for automation and high throughput, but shackling is

Figure 2 Water bath stunning of spent hens demonstrating entry to the water bath (a) and exit in an unconscious tonic state (b) image courtesy of Henny Reimert and Marien Gerritzen, Wageningen Livestock Research.

still done manually. While electrical water bath stunning can, in theory, result in immediate loss of consciousness, several major welfare concerns are associated with this approach. These include the risk of ineffective stunning (caused by inadequate current), the painful and aversive nature of shackling, and the risk of pre-stun shocks (e.g. electrically 'live' water on the ramp leading to the water bath coming into contact with the bird's body or wing tips). Other methods of electrical stunning (head only and head to body) are available but are not widely used in commercial slaughter plants, and much less research on these is available.

3.1.1 Determination of electrical stunning parameters for poultry

Because the effectiveness of electrical stunning depends on an adequate delivery of electrical current to the brain, it is affected by electrical variables such as voltage, current, frequency and waveform (Blokhuis et al., 2004; Kranen et al., 1996; Novoa et al., 2019; Raj et al., 2006c). A large body of research has been devoted to determining appropriate parameters for effective electrical stunning of poultry, and previous reviews are available (Berg and Raj, 2015; Blokhuis et al., 2004; Nielsen et al., 2019a; Raj, 2003, 2006; Shields and Raj, 2010). Various studies have determined the effectiveness of stunning parameters at a range of combinations of current, frequency and waveform, most commonly using either EEG output (e.g. Prinz et al., 2012) or reflex responses (e.g. Girasole et al., 2016) as outcome measures. Most of these have been based on 50 Hz alternating current (AC) with a sinusoidal waveform. The findings have underpinned evidence-based policy, and the stipulation of minimum currents according to different frequency ranges in law (e.g. EU Council Regulation (EC) No 1099/2009) (EC, 2009) and global guidelines (OIE (World Animal Health Organization), 2019).

An electrical stun is considered to be effective if it induces unconsciousness rapidly (in less than 1 s EU 1099/2009) and this is sustained for a minimum of 45 s (Gregory and Wotton, 1990b; Raj, 2006). It has been argued that use of measures such as muscle tone or convulsions to determine whether a stun has been successful are problematic (Shields and Raj, 2010), with EEG monitoring considered to be the most informative. In commercial conditions EEG measurements are not possible and other indicators of insensibility (e.g. loss of eye reflexes) are useful (Erasmus et al., 2010c; Prinz et al., 2010b). Signs of consciousness (e.g. wing flapping, responses to painful stimuli, vocalisations) may also indicate an ineffective stun (Grilli et al., 2015; Hindle et al., 2010).

In early work, it was determined that a minimum current of 100 mA per bird was required to ensure supressed EEG associated with a loss of consciousness in broilers, and 120 mA when using a 50 Hz sine wave AC (Gregory et al., 1991; Gregory and Wotton, 1990b, 1991; 1994). Raj et al. (2006c) reported that effective electrical stunning of broilers with a minimum constant current of 100, 150 and 200 mA could be achieved with electrical frequencies of up to 200, 600 and 800 Hz, respectively. Girasole et al. (2016) showed, under slaughterhouse conditions, that the minimum current necessary to achieve effective stunning in 90% of birds was 150 mA for 200 Hz, 200 mA for 400 Hz and 250 mA for 600 Hz. These and other studies show that higher stunning frequencies require greater current intensities to provide an effective stun, and this is reflected in current EU legislation (see Table 1) where allowable frequency/current combinations are defined (and some are not permitted). It is recommended that sine wave AC of 600 Hz maximum is delivered with an average current of 100-400 mA, depending on the species and the frequency (Council Regulation (EC) No 1099/2009) (EC, 2009). Raj et al. (2006a,b) investigated the application of different electrical waveforms, and concluded that sine wave AC is more effective than pulsed direct current (DC), and that a pulse width of at least 30% of current cycle is necessary to reliably induce epileptiform activity. Additionally, some electrical parameters, for example, low-frequency pulsed DC, are undesirable because they may induce cardiac arrest without an EEG pattern that indicates unconsciousness (Raj et al., 2006a). Although cardiac arrest is not necessary for an effective electrical stun, it is considered to be desirable, as it ensures non-recovery (Farm Animal Welfare Council, 2009). The duration of exposure to current is crucial to ensure adequate stunning and

Table 1 Minimum average current and frequency combinations as defined in the EU Council Regulation (EC) No 1099/2009 (EC, 2009) and OIE Terrestrial Animal Health Code (2005) (OIE (World Animal Health Organization), 2019)

Frequency (Hz)	Chickens (layers and broilers) (mA)	Turkeys (mA)	Ducks and geese (mA)
<200	100	250	130
200-400	150	400	Not permitted
400-1500	200	400	Not permitted

maintenance of unconsciousness during bleeding. A minimum of 4 s exposure to sine wave AC at a maximum of 600 Hz with a current of 100–400 mA is required by law in the EU but in practice, exposure times are routinely longer than this, in an attempt to diminish the possibility that birds will miss the water bath entirely (EC, 2009).

The meat quality implications of different electrical stunning parameters are routinely assessed because an inherent conflict exists between effective electrical stunning and economically relevant carcass quality outcomes. It is well known that while lower frequency current increases the likelihood of an effective stun, it increases intense muscle contractions leading to carcass defects associated with downgrading (Wilkins et al., 1999). Recent work indicated that a frequency of 750 Hz and current intensity of 200 mA resulted in effective stunning without adverse carcass lesions (Girasole et al., 2015), but others have reported that application of high current at high frequency also results in carcass defects (Ali et al., 2007). The opposition between carcass quality and humane electrical stunning appears to be unavoidable. While allowable electrical current and frequency parameters are set by law in the EU, the desire to promote meat quality has led to the global use of less effective currents that do not protect bird welfare (Bilgili, 1999; Hindle et al., 2010). For example, in the United States, electrical stunning for broilers routinely involves low current (10–45 mA per bird), high frequency (350–500 Hz), pulsed DC, which is unlikely to result in effective stunning (Shields and Raj, 2010). There are no regulations in place to control specific electrical stunning settings in the United States, and indeed AVMA guidelines (2020) allow electrical stunning with pulsed DC, low current (25–45 mA/bird average), low voltage (10–25 V) and high frequency (500 Hz), despite scientific evidence that these parameters are not effective (Leary et al., 2020).

3.1.2 Likelihood of effective electrical stunning

Even though suitable electrical parameters for stunning poultry have been determined, there are numerous factors that affect the likelihood of any individual bird being effectively stunned in a water bath system, and hence the reliability of the overall approach is uncertain. A significant welfare issue is the risk of inadequate current delivery to some birds, resulting in conscious birds experiencing pain and fear during subsequent processes. As outlined above, inadequate current could be caused by unsuitable underlying settings (voltage, frequency and waveform), but even when settings are conducive to an effective stun, it is well known that the amount of current applied to individual birds in a water bath stunner varies widely (Sparrey et al., 1992) and requires monitoring and reactive operational adjustment (Humane Slaughter Association, 2016a). Only a small proportion of applied current may flow through the brain, with the majority being distributed across the body (Woolley et al., 1986a,b), and resistance will vary between individuals according to body composition and contact with the shackles. Poor electrical contact with the shackle (caused by dryness, corrosion and debris; (Sparrey et al., 1992; Woolley et al., 1986b)) is therefore a risk, and may be reduced by maintaining shackle condition, cleanliness and wetting (Perez-Palacios and Wotton, 2006; Sparrey et al., 1992).

Due to the complexity of multi-bird water bath stunners, an ongoing issue is the difficulty in determining the actual current received by each bird (Hindle et al., 2010). As outlined above, multiple causes combine to result in a system that is not able to compensate for individual variation and therefore reliably deliver adequate current to all the birds, even if the settings are correct. A further complicating factor is that the visual appearance of inadequately stunned birds can be indistinguishable from effective stunning, since seizures induced by lower current applications effectively cause electro-immobilisation while birds remain conscious (Schutt-Abraham et al., 1983; Shields and Raj, 2010). Although there are some effective, commercially viable means to determine whether a bird is effectively stunned, routine use of rapid line speeds (such as 8000 birds per hour) limits the possibility to check this on an individual basis, and to apply backup interventions (Blokhuis et al., 2004). This is of grave concern because given the very large numbers of birds involved in commercial slaughter, even small percentage failure rates in stunning translate to an appreciable number of birds experiencing severely negative welfare consequences. Prevention is therefore the best approach, and one attempt to overcome the issue of inadequate current delivery is the use of a constant-current stunner, which delivers a pre-set constant current using variable voltages (compensating for the electrical impedance of individual birds). Such equipment has been developed (Sparrey et al., 1993) and tested with positive results (Lambooij et al., 2010), and its implementation would significantly improve the reliability and welfare outcomes associated with electrical stunning of poultry.

There are a couple of scenarios when birds completely avoid being electrically stunned in the water bath system: when they are too small for the head to reach the water, or when a bird lifts its head to avoid contacting the water, therefore entering the neck cutter while conscious. Live birds may also enter the scald tank (a post-death process to allow defeathering) because they miss the neck cutter after being stunned (and recovery ensues) or when both stunning and neck cutting are inadequate (Shields and Raj, 2010). The number of animals not stunned at all is not clear since no reliable figures are available. To avoid the poor welfare outcomes associated with these situations, EU legislation requires manual backup neck cutting, but again fast line speeds make this challenging to actually employ (Nielsen et al., 2019a).

3.1.3 Shackling

Concerns about inadequate current leading to ineffective stunning notwithstanding, a major welfare issue associated with electrical water bath stunning is the shackling of conscious birds. There is good evidence that this process causes pain, based on nociceptive activation by the compression forces generated by the weight of a 2-4 kg bird on the small surface area of the leg in contact with the shackle (Gentle and Tilston, 2000). As such, the size of the metal slot in relation to the bird's leg is important, with male broilers at greater risk of increased leg compression compared to smaller females (Satterlee et al., 2000; Sparrey and Kettlewell, 1994). In addition, inversion (hanging upside down) is a potent cause of stress and fear (Bedanova et al., 2007; Korte et al., 1997; Shields and Raj, 2010) and most birds struggle and wing flap immediately following shackling (Kannan et al., 1997). For birds that have leg abnormalities, injuries sustained during catching or inherent fragility (e.g. lame broilers or osteoporotic spent laying hens), the welfare consequences of shackling are likely to be more severe (Butterworth, 1999). Lambooij and Hindle (2018) reported that shackling increased rates of broken bones in spent hens, with 4-20% of freshly broken bones associated with handling immediately prior to stunning. Apart from the issues directly associated with shackling itself, poor manual handling to deliver birds to the shackles can also result in stress, injury, carcase downgrading or even mortality (Knowles and Broom, 1990).

The ongoing welfare costs associated with shackling mean that minimising its duration is a priority. Sparrey and Kettlewell (1994) recommended an optimal shackling time of between 12 and 60 seconds, which allows birds to 'relax' (i.e. stop wing flapping) to ensure an effective stun (Gregory and Bell, 1987) while avoiding longer than necessary shackling durations. The durations actually applied depend on line speed and plant design, which are widely variable. EU regulations stipulate that the shackling durations for chickens must not exceed 1 min (EC, 2009), but the extent of enforcement is not clear. Plant design considerations are also relevant to bird welfare with regard to avoiding sharp curves, steep inclinations or sudden drops in the shackle line, all of which can increase the mechanical forces on the legs and hence pain and fear responses (Nielsen et al., 2019a). Good electrical contact between the bird and the shackle is necessary for effective stunning, which introduces a tension between the degree of leg compression and a desire to use tight-fitting shackles. Compliant shackles have been developed; these are designed to adapt to bird leg size, ensuring good contact while minimising compression (Anastasov and Wotton, 2012; Lines et al., 2012), but are currently not widely used (Humane Slaughter Association, 2016b). A further modification to improve welfare is the introduction of breast support – this can be an almost vertical surface against which birds lean while hanging or take the form of a horizontal conveyor belt which bears weight. Horizontal supports have been shown to decrease wing flapping and should reduce pain and fear, but these are not widely used and measures must be taken to ensure that the birds are not able to remove their legs from the shackles (Lines et al., 2011a, 2012).

Given the inherent welfare insults associated with inversion and leg compression that cannot be dissociated from shackling of conscious birds, the best mitigation is to avoid the process entirely. This has been the starting point for alternative methods of stunning such as CAS. When shackling is used, harms can be somewhat minimised by staff training on good handling technique, ensuring both legs are shackled, not shackling injured or very small birds, minimising shackling duration, plant design that involves smooth and level transit of the shackle line, and use of compliant/well-fitting shackles and/ or breast support to reduce weight bearing on the legs (Fig. 3).

3.1.4 Pre-stun shocks

A further concern related to water bath stunning is the risk of pre-stun shocks, whereby an electric shock is applied to the bird before loss of consciousness. This can happen when a part of the bird (usually the wing) touches electrified

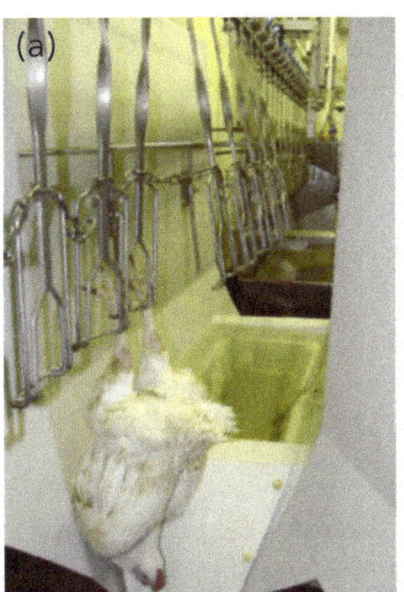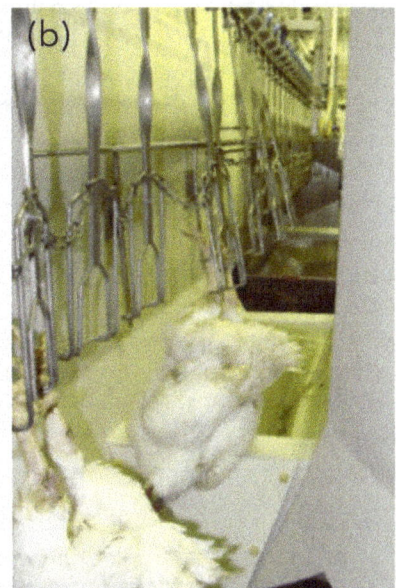

Figure 3 Broiler chicken approaching the water bath stunner (a) and then at the point of entry (b), where the angled ramp facilitates the swift and gentle swinging of the head into the electrified water in order to minimise the risk of pre-stun shocks. Image credit: Humane Slaughter Association (HSA). Reproduced with permission from the HSA from: Electrical Water bath Stunning of Poultry (Humane Slaughter Association, 2016b).

water before the head (Rao, 2014; Rao et al., 2013). The likelihood of a pre-stun shock is increased by wing flapping at the entrance to the bath or when water levels are too high, causing overflow to come into contact with birds at the entry ramp (Blokhuis et al., 2004). Such shocks are painful and the bird's reaction to them may cause further wing flapping, increasing the chance that the bird's head may miss the water bath completely (Humane Slaughter Association, 2016b; Nielsen et al., 2019a). Mitigation includes measures that promote a smooth and swift entry to the water bath (e.g. subdued lighting, breast support) and ensuring the water level in the bath is sufficient (birds should be immersed up to the base of their wings (Humane Slaughter Association, 2016b; Nielsen et al., 2019a)), while preventing overflow.

3.1.5 Head-only stunning

Application of current directly to the head (called 'head-only stunning') involves inversion of the bird into a cone or restraint by hand, and insertion of the head into a device with electrodes to deliver appropriate current (Hillebrand et al., 1996; Lambooij et al., 2010; Raj and O'Callaghan, 2004; Vogel et al., 2011). This approach, then, still involves stressful inversion, since restraint of conscious birds is required for proper positioning of electrodes. However this method is individual to each bird, increasing the chance of adequate current delivery and induction of unconsciousness, though variation resulting from the cleanliness and material of electrodes, feather coverage of head, dryness of skin, and other individual bird parameters remains (Nielsen et al., 2019a). Requirements for head-only electrical stunning equipment are regulated in the EU, being 240 mA for hens and broiler chickens (EC, 2009) and the same signs of unconsciousness are relevant as for water bath stunning. As with water bath stunning, head-only stunning is a simple stun and must be followed with a killing method. Although head-only methods are more likely to guarantee an effective stun, they are not suitable for high throughput and limited to small-scale operations. As such they have limited relevance to commercial poultry slaughter, although some research has investigated the feasibility of head-only stunning using a water bath (Lines et al., 2011b).

3.1.6 Novel electrical systems

Novel and emerging electrical stunning systems have been developed in an attempt to combine reliable, instantaneous stunning with welfare-friendly handling and restraint. With high throughput in mind, these have generally involved automated conveyor systems which mechanically locate and control the bird's head to allow the application of head-only stunning and/or head to body stun-killing. An 'upright' system was developed in the UK with initial promise (Tinker et al., 2005), but it has not been commercialised. Head-cloaca stunning has been attempted (Lambooij et al., 2012), which still involves shackling but uses the cloaca instead of the feet as the earthing point to improve electrical contact. It has shown promise in lab settings, but is not currently available. A novel system called 'Dutch Vision Solutions' applies individual head-only stunning. It has been found to be effective though some concerns were raised about prolonged stun to stick intervals and the reliability of backup systems, and it does not eliminate shackling of conscious birds (Gerritzen et al., 2015).

3.1.7 Conclusions

The welfare issues associated with electrical water bath stunning are considerable, and they are not easy to overcome. Despite this, electrical water bath systems remain the most commonly used stunning method for poultry worldwide. Even the use of constant current systems does not eliminate the significant harms associated with shackling, or the risk that some birds will miss the stunner completely. The move to alternatives, particularly in Europe, has been driven by a wish to eliminate these concerns. As early as 2004, EFSA recommended the phasing out of electrical methods due to their associated harms, with the Farm Animal Welfare Council (2009) endorsing this in 2009. In the latest EFSA report, there is a specific recommendation that birds should not be shackled while conscious, indirectly indicating that the majority of commercially used electrical stunning methods should cease (Nielsen et al., 2019a). While electrical stunning is still in use in the EU, an interim step has been to ensure it is heavily regulated. Insofar as risks to welfare can be mitigated, this is achieved by specific legislative requirements for water bath design, stipulation of specific currents and frequencies, requirements for water depth, a control box displaying current administered, capacity to adjust voltage, and appropriate back-up procedures for inadequate or missed stuns (Nielsen et al., 2019a). Appropriate inspection and enforcement is provided locally by each EU member state (EC, 2009). The lack of such oversight in other countries slaughtering billions of chickens with electrical stunning is a source of serious welfare concern.

3.2 Controlled atmosphere stunning

Controlled atmosphere stunning (CAS) involves exposing animals to modified gas environments, which induce a gradual loss of consciousness followed by a non-recovery state. The term CAS (as opposed to gas stunning) is used by the industry, partly to avoid negative public perceptions and historical connotations of 'gas chambers' (Berg and Raj, 2015). The term controlled atmosphere killing (CAK) is also used (interchangeably with CAS) to refer to this process, because although it is possible to induce a temporary loss of consciousness with gas exposure, all commercial systems are designed to stun-kill since birds tend to rapidly regain consciousness on re-exposure to air leading to a high risk of recovery on the slaughter line (Gerritzen et al., 2000, 2013; McKeegan et al., 2006, 2007; Poole and Fletcher, 1998; Raj, 1998; Raj and Gregory, 1990a). This is because birds have a unique, extremely efficient respiratory system which evolved to meet the demands of flapping flight (Fedde, 2012). Instead of alveoli (which are blind-ended structures in mammalian lungs where gas exchange takes place), parabronchi (a parallel series of tubes) are the primary unit of gas exchange in avian lungs (Fedde, 1998). Both inspiration and expiration require active muscle contractions and there is unidirectional air flow in a tail to head direction during both phases of ventilation which renews gas exchange more frequently than tidal (mammalian) ventilation (Powell and Scheid, 1989). As in mammals, breathing in birds is controlled by motor output from the brainstem and the rate and depth (tidal volume) of breathing is controlled by several reflexes designed to maintain blood gases within acceptable ranges. Central chemoreceptors in the brain are sensitive to changes in arterial carbon dioxide (CO_2) and pH, while arterial chemoreceptors (known as carotid bodies) respond to changes in arterial oxygen and pH and facilitate the ventilatory response to hypoxia (low oxygen) (Fedde, 2012; Powell and Scheid, 1989). In birds (but not in mammals), intrapulmonary (lung) chemoreceptors are stimulated by low CO_2 levels and inhibited by high CO_2 (Milsom et al., 2004). This inhibition of lung receptors when CO_2 is high stimulates an increase in ventilation (as in mammals but by a different mechanism).

Commercial systems vary in their design, but CAS is generally applied to birds in their transport containers or on conveyers, involving minimal or no handling of conscious birds. Crucially, shackling is performed after stunning when birds are in a non-recovery state. Correct application of CAS also involves timed exposure which ensures that all birds are stunned, and as such this approach eliminates two major welfare issues associated with electrical stunning – aversive handling and the risk of ineffective or absent stunning. However, unlike effective electrical stunning, induction of unconsciousness with controlled environments is not instantaneous, resulting in a transitional period during which negative experiences are possible. When assessing the welfare impact of CAS it therefore becomes vital to determine the time to loss of consciousness, bearing in mind that vigilance states ranging from alert to deep unconsciousness represent a continuum. As such, the transition from consciousness to unconsciousness should not be thought of as an 'on/ off' switch but a gradual process, especially with non-instantaneous stunning approaches such as CAS (Verhoeven et al., 2014). Although it remains challenging to characterise gradually changing brain states, EEG recording and analysis is widely agreed to be the best approach to determine the likelihood of consciousness and there have been attempts to define EEG spectral characteristics associated with different vigilance states in anaesthesia (alert, sedation, surgical plane, deep hypnotic state and brain death) (Benson et al., 2012; Gerritzen et al., 2004; McKeegan et al., 2007; Raj et al., 2006a; Sandercock et al., 2014) so that we can recognise them during stunning.

3.2.1 Types of controlled atmosphere stunning

CAS methods can be divided into five main categories which refer to the approach used to modify the environment, usually by displacing air in the system. These are anoxia (inert gases), hypercapnic anoxia (inert gases mixed with CO_2), hypercapnic hypoxia (gradual exposure to ultimately high levels of CO_2), hypercapnic hyperoxygenation (mixtures of inert gases, CO_2 and oxygen such that oxygen levels exceed those naturally found in air) and hypobaric hypoxia (low atmospheric pressure stunning (LAPS), in which birds are subject to gradual decompression). These have all been the basis of commercial CAS systems for poultry at some time, but most current systems employ hypercapnic hypoxia and/or hypercapnic hyperoxygenation, usually in multiple phases (Fig. 4). Systems with multiple steps are designed to first induce loss of consciousness, before application of a second phase (which may involve potentially aversive gas mixtures) to ensure non-recovery. The duration of the conscious phase is reported to be shortest for hypercapnic anoxia, followed by anoxia and hypercapnic hypoxia, and longest with hypercapnic hyperoxygenation (Johnson, 2014).

3.2.2 Behavioural responses to controlled atmosphere stunning

Behavioural responses can be used alongside brain measures to assess the stages of loss of consciousness and are the basis of welfare assessment during any stunning or killing method. Mandibulation is a behaviour which resembles a 'tasting' type movement of the bill and has been defined as 'Repetitive and rapid opening and closing of the bill, not associated with inspiration or exhalation' (Martin et al., 2016a,c). Mandibulation is seen in response to both anoxic and hypercapnic CAS mixtures, and LAPS (Abeyesinghe et al., 2007; Martin et al., 2016b). Its function and welfare impact are unclear. Headshaking is a rapid lateral head movement which has been previously associated with disorientation, discomfort and respiratory distress (Webster and Fletcher, 2001) as well as arousal or contexts demanding increased attention (such as the presentation of novel or disturbing stimuli (Hughes, 1983)). Nicol et al. (2011) concluded that head shaking may be a valid indicator of a less-preferred environment and high rates of head shaking are thought to indicate reduced welfare. Headshaking has been interpreted as an aversive reaction to CO_2, but it is also seen in response to stunning with inert gases and LAPS (Abeyesinghe et al., 2007; Johnson, 2014; Martin et al., 2016c; McKeegan et al., 2007). Ataxia describes the behavioural responses observed in the initial stages of animals

(a)

(b)

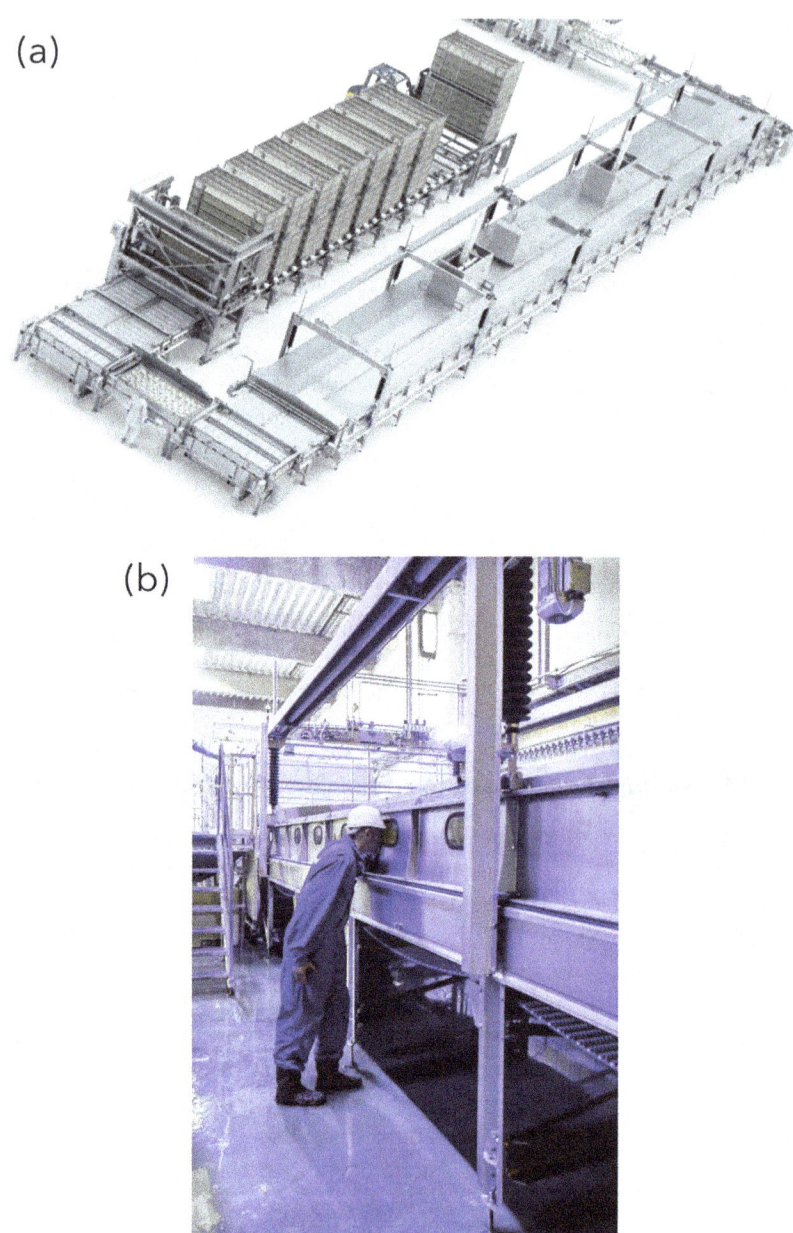

Figure 4 Marel's ATLAS live bird handling system and controlled atmosphere stunning (CAS) SmoothFlow (a) and the use of windows in the side of the system's hood allow broilers to be observed and monitored (b) images courtesy of Marel.

succumbing to gradual stunning methods such as CAS and it is characterised by progressive loss of voluntary coordination of muscle movement. It results in visible swaying of the body and/or head, attempts to adjust body posture and/or single wing flaps. It may be defined in poultry as 'Apparent dizziness, staggering, swaying of body and/or head, attempts to stand/sit or flaps wings to try and regain balance' (Martin et al., 2016a,b,c).

Rhythmic breathing with an open bill (distinguishable from thermoregulatory panting) is a hyperventilatory response to hypoxia (low oxygen). Deep inhalation (single large gasping breaths, often with neck extension) may also indicate dyspnoea (respiratory discomfort) (Mackie and McKeegan, 2016; Martin et al., 2016c). Open bill and deep breathing have been interpreted as indications of breathlessness in birds (Gerritzen et al., 2004), recently defined in mammals as a negative affective (emotional) experience relating to respiration with multiple qualities (Beausoleil and Mellor, 2015). In humans, these include respiratory effort, air hunger (increased urge to breath) and chest tightness (Beausoleil and Mellor, 2015), though it is not clear which (if any) of these apply to birds, which have a unique respiratory system. Dyspnoea-related behaviours probably reflect reduced welfare during stunning, and in humans there is evidence that air hunger is more unpleasant than pain (Banzett et al., 2008). Deep inhalation has been particularly associated with hyperventilation during CO_2 stunning, but it is also seen with inert gases (e.g. McKeegan et al., 2007) and after electrical stunning (Verhoeven et al., 2014). In CO_2 stunning, this behaviour continues after loss of posture indicating that consciousness is not required for its performance (Verhoeven et al., 2014). There is evidence that some dyspnoea occurs in all CAS stunning mixtures that have been investigated, including inert gases; however behavioural responses reflecting these have been variously described as 'gasping', 'deep breathing' and 'respiratory disruption' (e.g. Gerritzen et al., 2004; McKeegan et al., 2007) making direct comparison between behavioural categories more difficult.

Loss of posture is the inability of the animal to remain in an initial standing or sitting position and has been widely interpreted as a proxy for loss of consciousness (Authie et al., 2013; Gerritzen et al., 2004) as it reflects loss of muscle control. After loss of posture, convulsions consisting of neuromuscular spasms are observed, usually consisting of a clonic phase (characterized in chickens by vigorous wing flapping) and a tonic phase (characterized by rigidity and twitching of outstretched legs and wings), followed by final leg paddling motions (Erasmus et al., 2010a; Verhoeven et al., 2014). Wing flapping observed before loss of posture may be associated with purposeful escape attempts, indicating poor welfare (Nielsen et al., 2019a). Tonic and clonic convulsions sometimes alternate during CAS. The time at which convulsions cease has been used as an indicator of irreversible brain failure (Erasmus et al., 2010c). Convulsions are never seen before loss of posture, therefore their relevance to welfare is probably limited. However, it has been noted that when birds are killed in groups, it may be possible for birds that are still potentially conscious to be disturbed or even injured by other birds convulsing. Finally, birds will become motionless during CAS, which refers to a limp carcass with the bird being completely still including the cessation of visible breathing movements; it reflects complete and irreversible loss of muscle tone and is considered to confirm a non-recovery state. In this state the bird is no longer breathing and the brain is dead, but the heart (and possibly other body systems) will still be partially functioning (Abeyesinghe et al., 2007).

3.2.3 Welfare impacts of different controlled atmosphere stunning types

3.2.3.1 Inert gases

Immersion in inert gases (argon/nitrogen) causes minimal immediate visible and adverse reaction, since these gases cannot be sensed directly by the bird. Ataxia and loss of posture are apparent as the bird succumbs to lack of oxygen (which is sensed by arterial chemoreceptors (Fedde, 2012). While respiratory disruption and presumably a sensation of air hunger is induced by inert gases (in humans (Moosavi et al., 2003)), it is less common and overt compared to hypercapnic gas mixtures in birds (Coenen et al., 2009; Gerritzen et al., 2000; McKeegan et al., 2007). Anoxic CAS is associated with very vigorous convulsions which happen quite early in the process. While the welfare impact of these is unclear (McKeegan et al., 2007; Raj, 2009), they have implications for meat quality. EFSA (2004) opined that this sort of death (by 'pure' anoxia) is humane. However, the use of inert gases is practically problematic due to the difficulty of keeping oxygen below 2% (necessary to cause a non-recovery state in birds) (Raj et al., 1992) in a fast-throughput system. Argon is denser than air which helps to displace oxygen more effectively, but it is prohibitively expensive. As such, anoxic CAS is not common for commercial use.

3.2.3.2 Hypercapnia

Inhalation of CO_2 causes hypercapnia, which is an increase in CO_2 levels in the blood. This disrupts respiration and normal neuronal function, leading to unconsciousness and eventually to death. CO_2 becomes an anaesthetic gas (i.e. causes unconsciousness, and therefore could be used to stun reversibly) at concentrations of around 30% (Raj and Gregory, 1990b). The rate of induction and the depth of anaesthesia achieved depend on gas concentration and exposure time, with higher levels of CO_2 requiring shorter exposure times to induce unconsciousness (Raj and Gregory, 1990b). Irreversible stunning with

CO_2 is induced by hypoxia and acidification of cerebrospinal fluid, causing brain death (Blokhuis et al., 2004; Raj, 2009; Raj et al., 2006d; Terlouw et al., 2016).

Exposure to CO_2 is associated with two welfare concerns: possible activation of mucosal nociceptors, causing pain, and unpleasant respiratory sensations. Either or both of these may underlie aversion to CO_2, which has been observed in chickens (McKeegan, 2004; Raj and Gregory, 1991). Based on individual nociceptor recordings, the pain threshold for CO_2 exposure in birds is thought to be 45–50% (McKeegan, 2004) and this is reflected in the EU regulation, which stipulates a maximum level of 40% CO_2 to which conscious birds may be exposed (EC, 2009). No current CAS systems apply CO_2 concentrations higher than 40% to conscious birds. Pain, therefore, is not a major concern during CAS for poultry (unlike in pigs where conscious pigs are exposed to CO_2 concentrations in excess of 80% (Llonch et al., 2012; Rodríguez et al., 2008)).

Physiologically, inhalation of CO_2 causes three effects – (1) changes in blood CO_2 and pH sensed by central chemoreceptors in the brain; (2) changes in blood pH sensed by chemoreceptors in the arteries and (3) inhibition of intrapulmonary (lung) chemoreceptors (IPCs) (Fedde, 2012; Milsom et al., 2004). All of these drive a hypercapnic ventilatory response, resulting in prolonged inspiration, increased tidal volume (bigger breaths) and decreased breathing frequency. This is expressed behaviourally as visible respiratory disruption or dyspnoea. The thresholds for these responses (especially IPC inhibition) are low (<2%) (Milsom et al., 2004) relative to CO_2 concentrations that cause unconsciousness and death. The extent to which hypercapnic hypoxia multistage systems present a welfare issue depends on one's view of whether a negative affective (emotional) state is induced by CO_2 inhalation at sub-nociceptive concentrations. In humans, air hunger induced by inhalation of CO_2 is extremely unpleasant and stressful (Banzett et al., 2008), possibly additionally so since humans can understand the life-threatening implications of inadequate oxygen intake. Since birds have a very different respiratory system in terms of both structure and physiology, it is not clear whether it is valid to directly extrapolate the concept or mechanisms of air hunger from mammals to birds. In mammals, air hunger arises from a mismatch between the automatic motor command to breathe and the degree of lung inflation (Banzett et al., 2008; Beausoleil and Mellor, 2015). Birds have a fixed lung volume, unidirectional airflow and increased dead space (from air sacs), and so they may be better able to accommodate hyperventilation, reducing (or even eliminating) air hunger (Fedde, 2012). As such, behavioural manifestations of hyperventilation might have different welfare consequences in birds compared to mammals. However, if the ventilatory response induced by exposure to CO_2 is insufficient (i.e. when the reflex command exceeds normal functional respiratory capacity) it is certainly possible that this could lead to air hunger or a similarly negative sensation. In work on aversion to CAS mixtures, McKeegan et al. (2006) showed that broilers did not withdraw and continued to feed while showing behavioural evidence of dyspnoea in response to CO_2. This suggests that respiratory disruption may not in itself be potently aversive, but further study is needed to address this question. If dyspnoea is a potential cause for welfare concern as has been suggested (Beausoleil and Mellor, 2015), it is undesirable to use CAS systems that involve prolonged exposure to concentrations of CO_2 that cause dyspnoea, but are not anaesthetic. In other words, there is a tension between inducing unconsciousness very gradually (e.g. with multiple steps of slowly increasing CO_2 concentration) and allowing dyspnoea to continue over a long period. There is very little welfare research on multiphase hypercapnic hypoxic CAS, although it is widely used. In one study of a 5-stage CO_2 system, transitional EEG waveforms were apparent after 36 s, and the EEG was suppressed (indicating unconsciousness) at 60 s (Gerritzen et al., 2013), suggesting that conscious durations are not excessive and within the range reported for other CAS approaches (Johnson, 2014). Echoing the meat quality/welfare conflict with electrical stunning, multiphase hypercapnic hypoxia systems are popular since wing flapping is reduced, improving product quality (Johnson, 2014).

In hypercapnic hypoxic systems (particularly as CO_2 levels rise) or with hypercapnic anoxia, birds are also exposed to hypoxia as the relative amount of O_2 in the air decreases by displacement. Arterial chemoreceptors sense hypoxia leading to a hypoxic ventilatory response, also consisting of increased ventilation and tidal volume (Milsom et al., 2004). During hypercapnic hypoxia/ anoxia, therefore, both of these reflexes will be activated with possibly greater total dyspnoea (Beausoleil and Mellor, 2015; Steiner et al., 2019). In support of this, there is evidence that hyperoxgenation of hypercapnic gas mixtures causes a smoother induction to unconsciousness (as with anaesthetics), but the stun takes longer since the birds succumb only to CO_2 as an anaesthetic, not hypercapnic hypoxia (Johnson, 2014; McKeegan et al., 2007). Hyperoxygenated hypercapnia must be followed by a euthanasia phase (high CO_2) since birds will not be killed by such gas mixtures (McKeegan et al., 2007). This approach initially came from work in rats, whereby it was noticed that aversion to CO_2 was slightly reduced by supplemental oxygen, and that induction to unconsciousness was smoother (Coenen et al., 1995; Kirkden et al., 2008). However, it should be noted that the presence of supplemental oxygen does not reduce visible respiratory responses to CAS mixtures containing CO_2 (e.g. open bill breathing and deep inhalation) (McKeegan et al., 2007)). Whether hyperoxygenation reduces any sensation of 'air hunger' is unknown.

3.2.4 Low atmospheric pressure stunning

Low atmospheric pressure stunning (LAPS) involves placing birds in a sealed chamber (Fig. 5) that is gradually decompressed, leading to loss of consciousness and a non-recovery state (McKeegan et al., 2013b; Mackie and McKeegan, 2016; Martin et al., 2016a,b,c, 2019a). Decompression results in the reduction of the ambient atmospheric pressure and a proportional decrease in partial pressure of oxygen, leading to hypobaric hypoxia which is characterised by progressive loss of cognitive and psychomotor skills followed by loss of consciousness (Bouwsema and Lines, 2019; Holloway and Pritchard, 2017). As with other CAS methods, there is no pre-stun handling and birds are stunned in their transport crates. The original concept was developed in the United States in the 1990s following self-reports of a lack of pain or stress during hypobaric hypoxia in humans undergoing high-altitude pilot training (Gradwell, 2016; Gradwell and Macmillan, 2016; US Department of Transportation: Federal Aviation Administration, 2008). Following a positive opinion from EFSA (More et al., 2017), LAPS became legal in the EU for broilers up to 4 kg in May 2018 (Commission Implementing Regulation (EU) 2018/723 amending Annexes I and II to Council Regulation (EC) No 1099/2009 (EC, 2009)). As yet, it has not had significant uptake commercially.

LAPS is applied through a family of six decompression curves (Holloway and Pritchard, 2017; Martin et al., 2016c), which allow the system to automatically

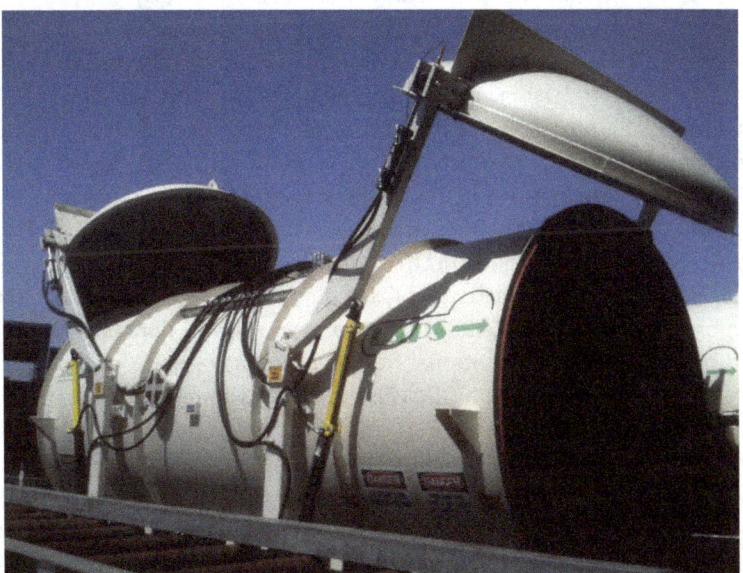

Figure 5 A chamber forming part of a low atmospheric pressure stunning (LAPS) system. Image courtesy of Randolph Cheek, TechnoCatch LLC.

adjust to variations in ambient temperature (−4°C-30°C) and associated water vapour pressure alterations (Wideman et al., 2013), in order to compensate for the deviation of different gas densities and maintain the same hypoxic effect. These adjustments appear to be effective, albeit resulting in small differences in behavioural latencies (Mackie and McKeegan, 2016; Martin et al., 2016c). In the 280 s LAPS cycle, the most rapid pressure change occurs in the first 67 s, averaging −1.41 kPa s⁻¹, before a second phase in which the rate of decompression is greatly reduced (68-180 s), averaging −0.10 kPa s⁻¹ (EC, 2009; Holloway and Pritchard, 2017). The final phase, referred to as the 'hold phase' involves a further minor reduction in atmospheric pressure in the final 100 s of the cycle; pressure is then maintained at around 19-20 kPa to ensure non-recovery. For comparison, atmospheric pressure at sea level is approximately 101 kPa and 33 kPa at the summit of Mount Everest (Newcomb, 2010). During the hold phase, the partial pressure of oxygen is approximately 2.7 kPa, which is equivalent to approximately 3.77% of atmospheric oxygen. This is higher than the recommended 2% atmospheric oxygen when using CAS with inert gases (Abeyesinghe et al., 2007; Blokhuis et al., 2004; McKeegan et al., 2013a), but it is sufficient to cause brain death and non-recovery (Martin et al., 2016b,c; McKeegan et al., 2013a). Decompression during LAPS also causes a reduction in temperature (approximately −4°C) and humidity (−50%) inside the chamber (Holloway and Pritchard, 2017).

The welfare concerns specifically associated with LAPS are air hunger responses to hypoxia and potential gas expansion in body cavities leading to the risk of pain and/or barotrauma. Behavioural responses to LAPS are consistent and the sequence of behaviours exhibited closely resembles those seen in normobaric hypoxic environments (Mackie and McKeegan, 2016; Martin et al., 2016a,b,c). The time to loss of consciousness during LAPS (as determined by behavioural and EEG indicators) is approximately 60 s (Martin et al., 2016a,c), which is within the range of other CAS systems (Abeyesinghe et al., 2007; Gerritzen et al., 2013; McKeegan et al., 2007). Respiratory disruption is apparent in LAPS environments (Mackie and McKeegan, 2016; Martin et al., 2016a,b,c), but to a lesser extent (both in counts and proportions of birds) than in hypercapnic or anoxic CAS (McKeegan et al., 2007). Concerns relating to the expansion of gases in internal body cavities have been raised by EFSA (Authie et al., 2014; More et al., 2017), especially with regard to the digestive tract and the sinuses and ears. This reflects locations where pressure-related discomfort has been reported in humans, but more usually on recompression (Bouwsema and Lines, 2019). Determining whether air expansion is causing pain during LAPS is challenging. Thorough pathological examinations of birds following LAPS have shown that organs and cavities are intact (Martin et al., 2019a), but a lack of pathological evidence does not eliminate the possibility of pain or discomfort during decompression. Martin et al. (2016b) investigated by applying an analgesic intervention during LAPS and reported no findings which would suggest colic-like pain or ear barotrauma. Like anoxic CAS, LAPS is associated with severe convulsions after loss of posture, which can have negative meat-quality consequences (such as a higher prevalence of wing damage (Martin et al., 2019a; Vizzier-Thaxton et al., 2010)). The changes in temperature and relative humidity observed during LAPS may have the potential to reduce welfare, for example, dryness of the nasal mucosae (Nielsen et al., 2019a). Though given the limited period of consciousness, this challenge would be brief.

3.2.5 Regulations and practical application

In the EU, only certain approaches to CAS are permitted for the routine slaughter of chickens. These are: two or more phases of successive exposure of conscious birds to a gas mixture containing up to 40% of CO_2, followed (when animals have lost consciousness) by a higher concentration of CO_2; direct or progressive exposure of conscious animals to a gas mixture containing up to 40% of CO_2 associated with inert gases; direct or progressive exposure of conscious animals to an inert gas mixture such as argon or nitrogen leading to anoxia. For LAPS an initial phase of a reduction in pressure from 760 Torr (sea level) to 250 Torr within a minimum of 50 s is permitted, and in the second phase a further reduction to 160 Torr within 210 s (Commission Implementing Regulation (EU) 2018/723 amending Annexes I and II to Council Regulation (EC) No 1099/2009 (EC, 2009)).

As with electrical stunning, the effectiveness and welfare impact of CAS in practice depends partly on care with implementation and process control. Nielsen et al. (2019a) identified three key risk areas for gas-based CAS: exposure of birds to painful concentrations of CO_2, too short exposure time to ensure stunning or inadequate gas concentration to ensure stunning. Mitigation of these requires proper monitoring of gas concentrations to maintain gas concentrations within legal limits while ensuring adequate exposure. For LAPS, Nielsen et al. (2019a) identified the hazards as too fast decompression, expansion of gases in the body cavity and too short exposure time, all of which are integral to the decompression curve used and its control by operators. Ambient temperature and humidity are also relevant to CAS methods, directly so for LAPS (as evidenced by adjustment of decompression curves (Holloway and Pritchard, 2017)) and because when birds experience heat stress they pant to thermoregulate by evaporative heat loss. Heat stress induces disturbances in blood gases and acid-base balance (called hypocapnic alkalosis – an increase in blood pH due to decreased partial pressure of arterial CO_2 (Sandercock et al., 2001)). This phenomenon probably explains anecdotal variation in responses to CAS that seem to be related to ambient temperature conditions, but this has not been formally studied.

3.2.6 Conclusions

CAS in poultry was developed to overcome aversive bird handling and improve the reliability of stunning in high throughput systems, and these important welfare benefits apply regardless of the approach. While process control and monitoring remain important, with advanced engineering and considerable progress in systems that can generate highly replicable controlled environments, the identified welfare hazards are less of an issue with CAS than electrical stunning (Authie et al., 2013). This does however mean that CAS systems require considerable capital investment and adequate space for installation in processing plants, both of which are barriers for smaller facilities. The overall perception of regulators and welfare commentators is that CAS represents a high welfare approach to poultry slaughter (Blokhuis et al., 2004; Farm Animal Welfare Council, 2009). Nevertheless, since most CAS systems in use in Europe involve exposure to CO_2, the biggest outstanding welfare question is the extent to which hypercapnic exposure induces respiratory responses that cause distress to birds. Research priorities related to this issue were recently outlined for CO_2 killing more generally (Steiner et al., 2019) and include correlation of respiratory variables to neurophysiologic and endocrine welfare indicators, further aversion testing and use of pharmacologic interventions to relate observable respiratory patterns to affective states. As noted above, this has been attempted with analgesics to investigate LAPS (Martin et al., 2016b). Use of anxiolytics is likely to be more relevant to air hunger but so far has not been investigated in relation to CAS systems for poultry.

3.3 Mechanical methods

Mechanical methods of stunning and killing are not used routinely in large-scale poultry slaughter and are reserved as emergency 'back up' methods, or are used in small-scale slaughtering and on-farm slaughter. Mechanical methods cause unconsciousness and death via destruction of the brain, damage to the brain, concussion or cerebral ischemia (Martin et al., 2018a, 2019b). These methods are not suitable for high throughput and are applied to individual birds (Fig. 6). All require appropriate restraint.

Application of a captive bolt is allowed in EU legislation (EC, 2009) and the bolt can be penetrative or non-penetrative. Captive bolt is stipulated as a reversible method; however, in practice it is irreversible if correctly applied, due to brain destruction (Erasmus et al., 2010b; Martin et al., 2018a). This method tends to be reserved for larger poultry species or individuals for which manual cervical dislocation is not suitable (Sparrey et al., 2014). Effective captive bolt

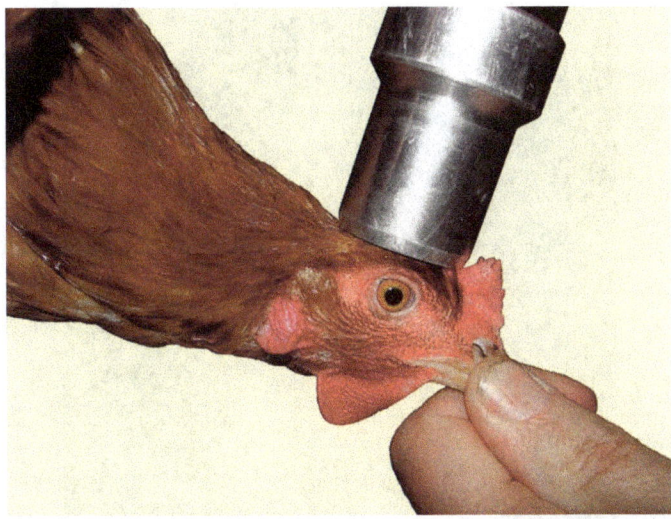

Figure 6 Captive bolt (concussive stunning) for chickens. Image credit: Humane Slaughter Association (HSA). Reproduced with permission from the HSA from: Practical Slaughter of Poultry (Humane Slaughter Association, 2016c).

application leads to immediate wing flapping and death can be confirmed by lack of cranial reflexes and absence of rhythmic breathing (Martin et al., 2018a; Raj and O'Callaghan, 2001; Woolcott et al., 2018). Recommended handling is that each bird is placed in a cone (which involves inversion) to access the head and contain wing flapping, and the forces applied (by cartridge, compressed air line pressure or a spring) should be appropriate for the species and size of birds (Nielsen et al., 2019b). As a last resort, EU regulations (EC, 2009) also allow the use of a percussive blow to the head (also known as blunt force trauma), described by Nielsen et al. (2019a,b) as 'holding a bird by its legs, placing its head on a hard surface and delivering a blow to the back of the head with a hard object' or 'holding the bird with both hands around its body and swinging the bird's head towards a hard, stable object such as the rim of a table'. Although there is evidence that such methods can be effective (Erasmus et al., 2010b; Raj and O'Callaghan, 2001), others have reported variable/limited success (Cors et al., 2015; Erasmus et al., 2010a), and there is clearly wide scope for operator error with severe welfare consequences. Better alternative methods are available, and as such use of percussive blows should be avoided.

For broilers of commercial market weight and spent hens, the most common method performed in emergency situations at slaughter is cervical dislocation (Sparrey et al., 2014) (Fig. 7). This can be performed manually (using just the hands of the operator) or mechanically (employing a tool). When performed optimally, cervical dislocation should cause death by cerebral ischaemia and extensive damage to the spinal cord and brainstem (Gregory

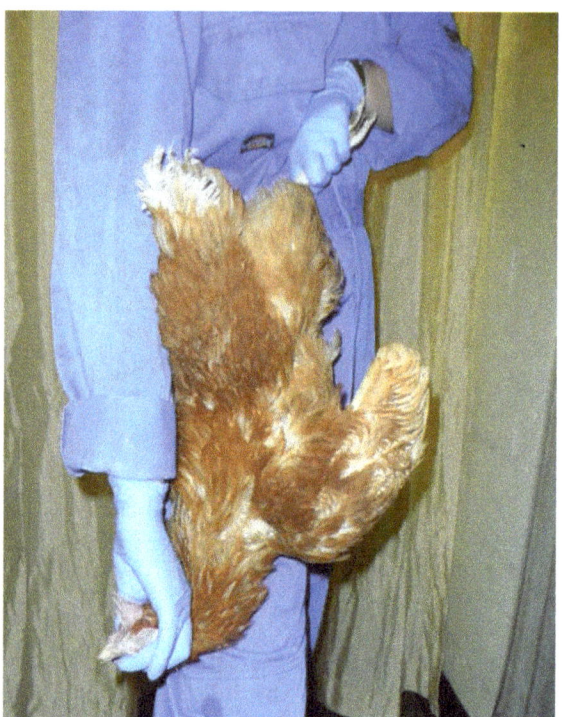

Figure 7 Demonstration of manual cervical dislocation in chickens. Image courtesy of Jessica Martin, University of Edinburgh.

and Wotton, 1990a; Martin et al., 2019b). Manual cervical dislocation requires training, experience and physical strength and involves firmly stretching and twisting the neck while tilting the head back in a continuous movement (Sparrey et al., 2014). Mechanical cervical dislocation employs devices that dislocate by stretching or crushing, normally after inversion of the conscious bird in a restraint cone. These include crushing at the first cervical vertebra with a pair of pliers such as 'Semark pliers' or the 'Burdizzo' (Erasmus et al., 2010a; Martin et al., 2017; Sparrey et al., 2014). In the EU, manual cervical dislocation can be performed on up to 70 birds/person per day (to minimise operator fatigue), although a recent study found no evidence that fatigue was apparent up to 100 birds (Martin et al., 2018a). Cervical dislocation can be performed manually on birds weighing up to 3 kg, but must be performed mechanically on birds weighing between 3 kg and 5 kg (no cervical dislocation is allowed on birds weighting more than 5 kg) (EC, 2009). A significant advantage of manual cervical dislocation is that it is available at any time without tools, which is important in emergency situations where there is ongoing suffering. However, suboptimal technique remains a concern with the potential to cause painful tissue damage and lack of separation of the brain and spinal cord resulting in delayed death. Mitigation involves rigorous staff training on correct application, which also applies to mechanical cervical dislocation and correct placement of captive bolt shots (Nielsen et al., 2019a,b).

The welfare implications of cervical dislocation, even when performed correctly, relate primarily to concern over how quickly consciousness is lost. Multiple studies have shown that neck crushing may not sever the carotid arteries or spinal cord and does not cause rapid loss of consciousness (Erasmus et al., 2010a; Gregory and Wotton, 1990a; Martin et al., 2018a,b). Recent publications have highlighted distinct differences in performance dependent on method (e.g. manual versus mechanical (Jacobs et al., 2019; Martin et al., 2016d, 2019b) and results have been conflicting regarding periods of reflex persistence or evidence of CNS processing, with wide inter-individual variation. For example, nictitating membrane persistence ranged from 2 s to 106 s in broilers and turkeys following cervical dislocation (Erasmus et al., 2010a; Martin et al., 2016d). Spectral analysis of EEG recordings on lightly anaesthetised broilers and hens subject to either manual or mechanical cervical dislocation indicated that unconsciousness was achieved on average 15 s following method application (Martin et al., 2019b). A limitation of several previous studies is that they conduct neurophysiological and/or reflex assessments when birds are in an anesthetised state, ranging from a general anaesthetic plane (Gregory and Wotton, 1990a) to lighter anaesthesia (Martin et al., 2019b; Woolcott et al., 2018), and lack of consistency between use of mechanical and manual methods. Recent efforts to enhance consistency have resulted in a novel tool, the 'Nex', which has been shown to improve the reliability and effectiveness of manual cervical dislocation (Martin et al., 2019b). This limits the relevance of earlier work to the skilled application of manual cervical dislocation in conscious birds and the lack of a comprehensive welfare assessment of this very common killing method represents a significant gap in knowledge.

3.4 Emerging methods

While still in the development stage, there are some methods of stunning that could have potential for future application to poultry. Application of gas mixtures in novel mediums such as high expansion gas-filled foam is a possibility. This approach has been assessed and is considered to be humane (Gerritzen and Sparrey, 2008; McKeegan et al., 2013a), but is currently used only for emergency killing for disease control. Foam is of interest because it facilitates the application of nitrogen (or other inert gases) to produce a rapid anoxic kill, by achieving low oxygen concentrations (< 1%) inside the foam, which are very difficult to achieve with free application of gases. Transcranial magnetic stimulation (TMS) has been trialled in broilers, and involves application of an intense magnetic field to the brain (Lambooij et al., 2011). Although this method appears to have the potential to induce a reversible state of unconsciousness, the period of insensibility was brief (up to 20 s) and further development work is required. Diathermic syncope involves directing microwave energy to the brain to cause a slight increase in brain temperature, which is associated with loss of consciousness (so-called hyperthermic syncope). This has been investigated in sheep and cattle with promising results (Small et al., 2013, 2019), but no research has been conducted into its application to poultry.

Before addition to the EU regulation, any novel method must protect animal welfare to a degree at least equivalent to that ensured by the existing permitted methods, and there is a significant burden of proof required with rigorous evidence standards (Nielsen et al., 2019a). For uptake to be viable, new methods must not be prohibitive economically, and of course ensure acceptable meat quality. For poultry, a requirement to allow the automation associated with high throughput is also a necessity.

3.5 Slaughter without stunning and religious requirements

3.5.1 Regulatory and religious requirements

The Islamic and Jewish faiths have specific requirements regarding the slaughter of animals for consumption. Both require slaughter methods that are based on scriptural rules, laid down in the Holy Quran and Torah, which require animals to be 'alive', healthy, and not injured at the time of slaughter (Fuseini et al., 2016). The Council of Europe's Convention for the Protection of Animals for Slaughter and EU Council Regulation 1099/2009 requires that animals are stunned before they are slaughtered, but both allow member states to apply derogations to allow meat production that meets the needs of religious communities. Despite this, several European countries do not allow slaughter without stunning including Sweden, Norway, Iceland, Denmark and Slovenia. Some countries stipulate that post-cut stunning must take place, including Austria, Estonia, Greece and Latvia (Global Legal Research Center, 2018).

Slaughter methods related to Islamic rules are termed halal (Arabic for permissible), while the Jewish equivalent is termed shechita (Velarde and Dalmau, 2018). While the Jewish faith rejects any form of stunning, some Islamic authorities allow the use of stunning, provided it does not kill the animals. This is usually defined by the presence of a heart beat at the time of neck cutting (Velarde et al., 2014), but there is no unanimous agreement about this definition or the acceptability of stunning among Islamic scholars (Chandia and Soon, 2018). In practice, the requirement for signs of life after stunning precludes the use of CAS systems (which stun-kill) and leaves electrical stunning (which is potentially reversible) as the only currently available method for halal stun slaughter of poultry. Because halal slaughter may involve reversible stunning or no stunning, 'non stun slaughter' is a more useful term than 'religious slaughter' in discussions about animal welfare consequences. Recent figures from the UK in one surveyed week (Defra, 2019) showed that for broilers 9.7% were non-stunned (9.4% halal and 0.3% shechita) and 12.6% were halal stunned, while for spent hens 1.4% were not stunned (all halal) and 6.4% were halal stunned.

3.5.2 Welfare impacts

The purpose of stunning is to render animals unconsciousness so that they do not experience exsanguination, so slaughter without stunning inevitably raises significant welfare concerns. These include aversive pre-slaughter handling, pain induced by the neck cut, distress during bleeding including aspiration of blood, and prolonged times to loss of consciousness (Anil, 2012). To minimize distress in both halal non-stun and shechita methods, the killing must be achieved by an uninterrupted cut from a sharp knife (that severs both jugular veins, carotid arteries, oesophagus and trachea) and by a competent individual using effective equipment (Fuseini et al., 2016; Velarde et al., 2014). This involves manual restraint and the presentation of the neck to a specialist slaughterman who makes the cut, and the bird is then placed into a bleed-out cone (Barnett et al., 2007).

Research is sparse on the welfare impacts of non-stun slaughter. Manual handling and inversion are potential sources of stress and injury for non-stun slaughter, and shackling may be used in electrical halal stunning with its concomitant welfare issues. There have been a few attempts to determine time to loss of consciousness following neck cutting in conscious chickens. Gregory and Wotton (1986) showed that 95% of the electrical-evoked activity in the brain was lost on average by 163 s after the cutting of both carotid arteries and by 349 s after cutting both jugular veins. However, they noted that birds were unlikely to be conscious throughout the entire period when there was measurable electrical activity in their brains. Barnett et al. (2007) reported that chickens undergoing shechita slaughter showed a minimal reaction to the neck cut, but 60% of them showed eye reflex responses 5 s after the neck cut. These were abolished by 15 s, which coincided with loss of muscle control. In a case study, Cranley (2017) described basic behavioural measures during non-stunned slaughter of 250 chickens, and reported that 225 chickens 'died at or before 60 s', but 25 chickens showed prolonged spontaneous behaviour for more than 90 s after the neck cut.

Electrical stunning can meet the needs of halal slaughter because it has the potential to deliver birds to the neck cutter alive and 'undamaged'. However, in systems employing a water bath, it is subject to the same limitations of aversive shackling and unreliability with regard to current delivery, and indeed more so since the stun needs to be demonstrably reversible and so cardiac arrest must not be induced. In a survey of religious slaughter spanning multiple species, Velarde et al. (2014) reported that 15% of the electrically stunned broilers assessed were breathing at the moment of neck cutting and 5% of them were breathing 30 s later. Although breathing does not directly relate to consciousness, it was suggested that the high percentage of rhythmic breathing observed after stunning indicates that the settings of the system were suboptimal. As discussed above, incorrect electrical parameters are likely to result in electro-immobilisation rather than unconsciousness with negative welfare consequences during neck cutting and bleeding. Wotton et al. (2014) suggested that application of 200 mA at 1000 Hz was sufficient to avoid cardiac arrest and so meet halal requirements, but indicators of' consciousness were not measured. Given that electrical stunning remains the only viable method for a reversible stun, novel automated head-only systems (such as 'Dutch Vision Solutions') are in development with the halal slaughter market in mind, and may provide opportunities to improve welfare by ensuring adequate current delivery on an individual basis.

3.5.3 Mitigation and conclusions

Various strands of evidence suggest that slaughter to meet religious needs is associated with poor welfare outcomes, even when stunning is used. In non-stun slaughter, opportunities to mitigate these are limited, but meticulous technique and care with handling are desirable, and EU regulations require that where animals are killed without stunning by a religious method the persons responsible for slaughtering must carry out systematic checks to ensure that animals do not present any signs of consciousness or sensibility before further processing (EC, 2009). Further development of electrical stunning systems to meet both welfare and religious needs is urgently required, bearing in mind that the size of these markets is growing (Defra, 2019; Fuseini et al., 2016; Velarde and Dalmau, 2018). Religious slaughter, especially without stunning, remains a controversial issue that is unlikely to be resolved due to the fundamental conflict it presents between animal welfare and human rights with regard to religious freedom (Anil, 2012).

4 Conclusions

The stunning and slaughter of chickens is a major commercial enterprise and methods applied must provide a balance between cost, feasibility and bird welfare outcomes. Both electrical stunning and controlled atmosphere methods have strengths and weaknesses in these regards. Electrical stunning has the important advantage of being theoretically instantaneous, but reliability is low in practice and it is associated with highly aversive handling of conscious birds. CAS resolves these issues, but potentially causes welfare harms during gradual induction of unconsciousness. As such, both methods are associated with variation in welfare outcomes based on process aims (in terms of the actual settings applied – electrical parameters or gas mixtures used) in addition to how well these settings are actually implemented in commercial use (i.e. the effectiveness of process control), especially under challenging high throughput conditions. For electrical stunning, suitable parameters to induce unconsciousness of sufficient duration are known, but there is an inherent tension between reliable stunning to protect welfare and poorer meat quality outcomes. In addition, there has been reluctance to take up novel equipment which would improve welfare as part of electrical stunning. The reasons for this are not clear, but are likely related to cost, challenges to practical implementation at high line speeds and in some cases, lack of funding for commercialisation. With CAS, the reliability of stunning is more assured, but uncertainty remains about the welfare implications of exposure to certain controlled atmospheres, especially sub-nociceptive concentrations of CO_2. This is a priority area for future research, given that systems which gradually expose birds to progressively increasing concentrations of CO_2 are widely used in Europe. An improved understanding of the extent to which avian respiratory systems produce breathlessness, air hunger and distress in hypercapnic and hypoxic environments is needed to assess likely aversion to CAS processes. Welfare outcomes are likely to vary in extent and mechanism depending on gas mixture, and comparisons of hypoxic dyspnoea versus hypercapnic dyspnoea, for example, are needed to underpin the development of improved CAS systems. Following repeated calls to end the use of electrical stunning in the EU, its future use is likely to be confined to small-scale operations for halal markets, which require reversible stunning (which CAS cannot provide). For such a system to be welfare-friendly, it must achieve non-aversive handling and reliable delivery of sufficient current to each bird, and novel systems are making progress towards this, encouraging optimism. It is probably impossible to accomplish stunning and slaughter (especially on a large scale) without imposing some welfare costs on the birds. However, striving to minimise harms to the billions of animals involved is a vital aim, and opportunities to improve welfare at chicken slaughter arise from continued research into better methods, and wider adoption of approaches already known to have acceptable welfare costs. On a global scale and with current knowledge, this leads to a strong recommendation to replace electrical stunning with CAS.

5 Where to look for further information

- Authie, E., Berg, C., Bøtner, A., Browman, H., Capua, I., Koeijer, A. de, Depner, K., Domingo, M., Edwards, S., Fourichon, C., Koenen, F., More, S.,

Raj, M., Shivonen, L., Spoolder, H., Stegeman, J. A., Thulke, H.-H., Velarde, A., Vågsholm, I., Willeberg, P., Zientara, S., 2013. Scientific Opinion on monitoring procedures at slaughterhouses for poultry. *EFSA J.* https://doi .org/10.2903/j.efsa.2013.3521.

- Berg, C., Raj, M., 2015. A review of different stunning methods for Poultry–animal welfare aspects (stunning methods for poultry). *Animals* 5, 1207-1219. https://doi.org/10.3390/ani5040407.
- Bøtner, A., Broom, D., Doherr, M. G., Domingo, M., Hartung, J., Keeling, L., Koenen, F., More, S., Morton, D., Oltenacu, P., Salati, F., Salman, M., Sanaa, M., Sharp, J. M., Stegeman, J. A., Szücs, E., Thulke, H.-H., Vannier, P., Webster, J., Wierup, M., 2012. Scientific Opinion on the electrical requirements for waterbath stunning equipment applicable for poultry. *EFSA J.* 10. https://doi.org/10.2903/j.efsa.2012.2757.
- Johnson, C. L. C., 2014. A review of bird welfare during controlled atmosphere and electrical water-bath stunning. *J. Am. Vet. Med. Assoc.* 245, 60-68. https://doi.org/10.2460/javma.245.1.60.
- Leary, S., Underwood, W., Anthony, R., Cartner, S., Grandin, T., Greenacre, C., Gwaltney-Brant, S., McCrackin, M. A., Meyer, R., Miller, D., Shearer, J., Turner, T., Yanong, R., Johnson, C. L., Patterson-Kane, E., 2020. *AVMA Guidelines for the Euthanasia of Animals: 2020 Edition*.
- Nielsen, S. S., Alvarez, J., Bicout, D. J., Calistri, P., Depner, K., Drewe, J. A., Garin-Bastuji, B., Gonzales Rojas, J. L., Gortázar Schmidt, C., Miranda Chueca, M. Á., Roberts, H. C., Sihvonen, L. H., Spoolder, H., Stahl, K., Velarde Calvo, A., Viltrop, A., Winckler, C., Candiani, D., Fabris, C., Van der Stede, Y., Michel, V., 2019a. Slaughter of animals: poultry. *EFSA J.* 17. https://doi.org /10.2903/j.efsa.2019.5849.
- Steiner, A. R., Flammer, S. A., Beausoleil, N. J., Berg, C., Bettschart-Wolfensberger, R., Pinillos, R. G., Golledge, H. D., Marahrens, M., Meyer, R., Schnitzer, T., Toscano, M. J., Turner, P. V., Weary, D. M., Gent, T. C., 2019. Humanely ending the life of animals: research priorities to identify alternatives to carbon dioxide. *Animals* 9, 1-25. https://doi.org/10.3390/ ani9110911.
- Velarde, A., Dalmau, A., 2018. Slaughter without stunning, in: *Advances in Agricultural Animal Welfare*. Woodhead Publishing, Duxford, UK, pp. 221-240.

6 References

Abeyesinghe, S., McKeegan, D., McLeman, M., Lowe, J., Demmers, T., White, R., Kranen, R., Van Bemmel, H., Lankhaar, J., Wathes, C., Abeyesinghe, S., McLeman, M., Lowe, J., Demmers, T., White, R., Kranen, R., Van Bemmel, H., Lankhaar, J. and Wathes, C. 2007. Controlled atmosphere stunning of broiler chickens. II. Effects on behaviour, physiology and meat quality in a commercial processing plant. *Br. Poult. Sci.* 48(4), 430–442. https://doi.org/10.1080/00071660701543097.

Ali, A. S., Lawson, M. A., Tauson, A. H., Fris Jensen, J. and Chwalibog, A. 2007. Influence of electrical stunning voltages on bleed out and carcass quality in slaughtered broiler chickens. *Arch. Geflugelkd.* 71, 35–40.

Anastasov, M. I. and Wotton, S. B. 2012. Survey of the incidence of post-stun behavioural reflexes in electrically stunned broilers in commercial conditions and the relationship of their incidence with the applied water-bath electrical parameters. *Anim. Welf.* 21(2), 247–256. https://doi.org/10.7120/09627286.21.2.247.

Anil, M. H. 2012. Religious slaughter: A current controversial animal welfare issue. *Anim. Front.* 2(3), 64–67. https://doi.org/10.2527/af.2012-0051.

Authie, E., Berg, C., Bøtner, A., Browman, H., Capua, I., Koeijer, A., Depner, K., Domingo, M., Edwards, S., Fourichon, C., Koenen, F., More, S., Raj, M., Sihvonen, L., Spoolder, H., Stegeman, J., Thulke, H., Vågsholm, I., Velarde, A., Willeberg, P. and Zientara, S. 2014. Scientific Opinion on the use of low atmosphere pressure system (LAPS) for stunning poultry. *Eur. Food Saf. Auth J.* 12, 1–27.

Authie, E., Berg, C., Bøtner, A., Browman, H., Capua, I., Koeijer, A. de, Depner, K., Domingo, M., Edwards, S., Fourichon, C., Koenen, F., More, S., Raj, M., Shivonen, L., Spoolder, H., Stegeman, J. A., Thulke, H.-H., Velarde, A., Vågsholm, I., Willeberg, P. and Zientara, S. 2013. Scientific Opinion on monitoring procedures at slaughterhouses for poultry. *EFSA J.* 11(12) https://doi.org/10.2903/j.efsa.2013.3521.

Banzett, R. B., Pedersen, S. H., Schwartzstein, R. M. and Lansing, R. W. 2008. The affective dimension of laboratory dyspnea: air hunger is more unpleasant than work/effort. *Am. J. Respir. Crit. Care Med.* 177(12), 1384–1390. https://doi.org/10.1164/rccm.2 00711-1675OC.

Barnett, J. L., Cronin, G. M. and Scott, P. C. 2007. Behavioural responses of poultry during kosher slaughter and their implications for the birds' welfare. *Vet. Rec.* 160(2), 45–49. https://doi.org/10.1136/vr.160.2.45.

Bayliss, P. A. and Hinton, M. H. 1990. Transportation of broilers with special reference to mortality rates. *Appl. Anim. Behav. Sci.* 28(1–2), 93–118. https://doi.org/10.1016/0 168-1591(90)90048-I.

Beausoleil, N. J. and Mellor, D. J. 2015. Introducing breathlessness as a significant animal welfare issue. *N. Z. Vet. J.* 63(1), 44–51. https://doi.org/10.1080/00480169.2014.94 0410.

Bedanova, I., Voslarova, E., Chloupek, P., Pistekova, V., Suchy, P., Blahova, J., Dobsikova, R. and Vecerek, V. 2007. Stress in broilers resulting from shackling. *Poult. Sci.* 86(6), 1065–1069. https://doi.org/10.1093/PS/86.6.1065.

Benson, E. R., Alphin, R. L., Rankin, M. K., Caputo, M. P., Kinney, C. A. and Johnson, A. L. 2012. Evaluation of EEG based determination of unconsciousness vs. loss of posture in broilers. *Res. Vet. Sci.* 93(2), 960–964. https://doi.org/10.1016/j.rvsc.2011.12.008.

Berg, C. and Raj, M. 2015. A review of different stunning methods for Poultry–animal welfare aspects (stunning methods for poultry). *Animals* 5(4), 1207–1219. https://doi .org/10.3390/ani5040407.

Bilgili, S. F. 1999. Recent advances in electrical stunning. *Poult. Sci.* 78(2), 282–286. https:// doi.org/10.1093/PS/78.2.282.

Blokhuis, H. J., Roth, B., Holst, S., Kestin, S., Raj, M., Terlouw, C., Calvo, A. V. and Von Wenzlawowicz, M. 2004. Opinion of the Scientific Panel on Animal Health and Welfare on a request from the Commission related to welfare aspects of the main systems of stunning and killing the main commercial species of animals. *EFSA J.* 45, 1–29.

Bøtner, A., Broom, D., Doherr, M., Domingo, M., Hartung, J., Keeling, L., Koenen, F., More, S., Morton, D., Oltenacu, P., Osterhaus, A., Salati, F., Salman, M., Sanaa, M., Sharp, M., Stegeman, J., Szücs, E., Thulke, H., Vannier, P., Webster, J. and Wierup, M. 2011. Scientific opinion concerning the welfare of animals during transport. *EFSA J.* 9(1), 1–125. https://doi.org/10.2903/j.efsa.2011.1966.

Bøtner, A., Broom, D., Doherr, M. G., Domingo, M., Hartung, J., Keeling, L., Koenen, F., More, S., Morton, D., Oltenacu, P., Salati, F., Salman, M., Sanaa, M., Sharp, J. M., Stegeman, J. A., Szücs, E., Thulke, H.-H., Vannier, P., Webster, J. and Wierup, M. 2012. Scientific Opinion on the electrical requirements for waterbath stunning equipment applicable for poultry. *EFSA J.* 10(6). https://doi.org/10.2903/j.efsa.2012.2757.

Bouwsema, J. and Lines, J. 2019. Could low atmospheric pressure stunning (LAPS) be suitable for pig slaughter? A review of available information. *Anim. Welf.* 28(4), 421–432. https://doi.org/10.7120/09627286.28.4.421.

Butterworth, A. 1999. Infectious components of broiler lameness: a review. *World's Poult. Sci. J.* 55(4), 327–352. https://doi.org/10.1079/WPS19990024.

Caffrey, N. P., Dohoo, I. R. and Cockram, M. S. 2017. Factors affecting mortality risk during transportation of broiler chickens for slaughter in Atlantic Canada. *Prev. Vet. Med.* 147, 199–208. https://doi.org/10.1016/j.prevetmed.2017.09.011.

Chandia, M. and Soon, J. M. 2018. The variations in religious and legal understandings on halal slaughter. *Br. Food J.* 120(3), 714–730. https://doi.org/10.1108/BFJ-03-2017-0129.

Cockram, M. S., Dulal, K. J., Mohamed, R. A. and Revie, C. W. 2019. Risk factors for bruising and mortality of broilers during manual handling, module loading, transport, and lairage. *Can. J. Anim. Sci.* 99(1), 50–65. https://doi.org/10.1139/cjas-2018-0032.

Coenen, A. M., Lankhaar, J., Lowe, J. C. and McKeegan, D. E. 2009. Remote monitoring of electroencephalogram, electrocardiogram, and behavior during controlled atmosphere stunning in broilers: implications for welfare. *Poult. Sci.* 88(1), 10–19. https://doi.org/10.3382/ps.2008-00120.

Coenen, A. M., Drinkenburg, W. H., Hoenderken, R. and van Luijtelaar, E. L. 1995. Carbon dioxide euthanasia in rats: oxygen supplementation minimizes signs of agitation and asphyxia. *Lab. Anim.* 29(3), 262–268. https://doi.org/10.1258/002367795781088289.

Cors, J. C., Gruber, A. D., Günther, R., Meyer-Kühling, B., Esser, K. H. and Rautenschlein, S. 2015. Electroencephalographic evaluation of the effectiveness of blunt trauma to induce loss of consciousness for on-farm killing of chickens and turkeys. *Poult. Sci.* 94(2), 147–155. https://doi.org/10.3382/PS/PEU038.

Cranley, J. 2017. Death and prolonged survival in nonstunned poultry: A case study. *J. Vet. Behav. Clin. Appl. Res.* 18, 92–95. https://doi.org/10.1016/j.jveb.2016.09.005.

Defra 2019. *Results of the 2018 FSA Survey into Slaughter Methods in England and Wales.* Defra, London, UK.

Devos, G., Moons, C. P. H. and Houf, K. 2018. Diversity, not uniformity: slaughter and electrical waterbath stunning procedures in Belgian slaughterhouses. *Poult. Sci.* 97(9), 3369–3379. https://doi.org/10.3382/ps/pey181.

Ekstrand, C. 1998. An observational cohort study of the effects of catching method on carcase rejection rates in broilers. *Anim. Welf.* 7, 87–96.

Erasmus, M. A., Lawlis, P., Duncan, I. J. and Widowski, T. M. 2010a. Using time to insensibility and estimated time of death to evaluate a nonpenetrating captive bolt, cervical dislocation, and blunt trauma for on-farm killing of turkeys. *Poult. Sci.* 89(7), 1345–1354. https://doi.org/10.3382/ps.2009-00445.

Erasmus, M. A., Turner, P. V., Nykamp, S. G. and Widowski, T. M. 2010b. Brain and skull lesions resulting from use of percussive bolt, cervical dislocation by stretching, cervical dislocation by crushing and blunt trauma in turkeys. *Vet. Rec.* 167(22), 850–858. https://doi.org/10.1136/vr.c5051.

Erasmus, M. A., Turner, P. V. and Widowski, T. M. 2010c. Measures of insensibility used to determine effective stunning and killing of poultry. *J. Appl. Poult. Res.* 19(3), 288–298. https://doi.org/10.3382/japr.2009-00103.

European Commission 2009. Council regulation (EC) No 1099/2009 of 24 September 2009 on the protection of animals at the time of killing. OJ L 303, 18.11.2009, p. 1–30. Available at: ELI: http://data.europa.eu/eli/reg/2009/1099/oj.

European Commission 2017. Preparation of best practices on the protection of animals at the time of killing. Final Report [WWW Document]. https://doi.org/10.2875/15243.

European Council 1979. European Convention for the protection of animals for slaughter [WWW Document]. Eur. Treaty Ser. Available at: https://www.coe.int/en/web/conventions/full-list/-/conventions/rms/0900001680077d98.

European Food Safety Authority (EFSA). (2004). Opinion of the Scientific Panel on Animal Health and Welfare (AHAW) on a request from the Commission related to welfare aspects of the main systems of stunning and killing the main commercial species of animals. EFSA Journal, 2(7), 45. doi.org/10.2903/j.efsa.2004.45.

European Union 2017. Brexit: farm animal welfare, Chapter 2: Maintaining standards [WWW Document]. Rep. Sess.

Farm Animal Welfare Council 2009. *Report on the Welfare of Farmed Animals at Slaugther or Killing*. Crown Publishing, London, UK.

Fedde, M. 2012. Respiration. In: Sturkie, P. (Ed), *Avian Physiology*. Springer-Verlag New York Inc., New York, pp. 191–221. https://doi.org/10.1007/978-1-4612-4862-0.

Fedde, M. R. 1998. Relationship of structure and function of the avian respiratory system to disease susceptibility. *Poult. Sci.* 77(8), 1130–1138. https://doi.org/10.1093/PS/77.8.1130.

Fetrow, J., Nordlund, K. V. and Norman, H. D. 2006. Invited review: culling: nomenclature, definitions, and recommendations. *J. Dairy Sci.* 89(6), 1896–1905 https://doi.org/10.3168/jds.S0022-0302(06)72257-3.

Food and Agriculture Organization of the United Nations (FAO) 2020. Livestock primary, producing animals/slaughtered [WWW Document]. *FAO Stat.* Available at: http://www.fao.org/faostat/en/#data/QL (Accessed 28 February 2020).

Fuseini, A., Knowles, T. G., Hadley, P. J. and Wotton, S. B. 2016. Halal stunning and slaughter: criteria for the assessment of dead animals. *Meat Sci.* 119, 132–137. https://doi.org/10.1016/J.MEATSCI.2016.04.033.

Gentle, M. J. and Tilston, V. L. 2000. Nociceptors in the legs of poultry: implications for potential pain in pre-slaughter shackling. *Anim. Welf.* 9, 227–236.

Gerritzen, M. A., Lambooij, B., Reimert, H., Stegeman, A. and Spruijt, B. 2004. On-farm euthanasia of broiler chickens: effects of different gas mixtures on behavior and brain activity. *Poult. Sci.* 83(8), 1294–1301. https://doi.org/10.1093/ps/83.8.1294.

Gerritzen, M. and Sparrey, J. 2008. A pilot study to assess whether high expansion CO2-enriched foam is acceptable for on-farm emergency killing of poultry.

Gerritzen, M., van Hattum, T. and Reimert, H. 2015. Efficacy of the Dutch Vision high-low electrical head-only poultry stunner. Livestock Research Report. Wageningen, The Netherlands. https://doi.org/10.13140/RG.2.1.2174.3767.

Gerritzen, M. A., Lambooij, E., Hillebrand, S. J. W., Lankhaar, J. A. C. and Pieterse, C. 2000. Behavioral responses of broilers to different gaseous atmospheres. Poult. Sci. 79(6), 928–933. https://doi.org/10.1093/ps/79.6.928.

Gerritzen, M. A., Reimert, H. G. M., Hindle, V. A., Verhoeven, M. T. W. and Veerkamp, W. B. 2013. Multistage carbon dioxide gas stunning of broilers. Poult. Sci. 92(1), 41–50. https://doi.org/10.3382/ps.2012-02551.

Girasole, M., Chirollo, C., Ceruso, M., Vollano, L., Chianese, A. and Cortesi, M. L. 2015. Optimization of stunning electrical parameters to improve animal welfare in a poultry slaughterhouse. Ital. J. Food Saf. 4(3), 175–178. https://doi.org/10.4081/ijfs.2015.4576.

Girasole, M., Marrone, R., Anastasio, A., Chianese, A., Mercogliano, R. and Cortesi, M. L. 2016. Effect of electrical water bath stunning on physical reflexes of broilers: evaluation of stunning efficacy under field conditions. Poult. Sci. 95(5), 1205–1210. https://doi.org/10.3382/PS/PEW017.

Global Legal Research Center 2018. Legal restrictions on religious slaughter in Europe [WWW Document]. Law Libr. Congress. Available at: https://www.loc.gov/law/help/religious-slaughter/europe.php (Accessed 3 February 2020).

Gradwell, D. 2016. Hypoxia and hyperventilation. In: Gradwell, D. and Rainford, D. (Eds), Ernsting's Aviation and Space Medicine. CRC Press, Boca Raton, FL, p. 904.

Gradwell, D. and Macmillan, A. 2016. Oxygen systems, pressure cabin and clothing. In: Gradwell, D. and Rainford, D. (Eds), Ernsting's Aviation and Space Medicine. CRC Press, Boca Raton, FL.

Gregory, N. G. 2008. Animal welfare at markets and during transport and slaughter. Meat Sci. 80(1), 2–11. https://doi.org/10.1016/j.meatsci.2008.05.019.

Gregory, N. G. and Bell, J. C. 1987. Duration of wing flapping in chickens shackled before slaughter. Vet. Rec. 121(24), 567–569. https://doi.org/10.1136/VR.121.24.567.

Gregory, N. G., Wilkins, L. J. and Wotton, S. B. 1991. Effect of electrical stunning frequency on ventricular fibrillation, downgrading and broken bones in broilers, hens and quails. Br. Vet. J. 147(1), 71–77. https://doi.org/10.1016/0007-1935(91)90069-Y.

Gregory, N. G. and Wotton, S. B. 1986. Effect of slaughter on the spontaneous and evoked activity of the brain. Br. Poult. Sci. 27(2), 195–205. https://doi.org/10.1080/00071668608416872.

Gregory, N. G. and Wotton, S. B. 1987. Effect of electrical stunning on the electroencephalogram in chickens. Br. Vet. J. 143(2), 175–183. https://doi.org/10.1016/0007-1935(87)90009-1.

Gregory, N. G. and Wotton, S. B. 1990a. Comparison of neck dislocation and percussion of the head on visual evoked responses in the chicken's brain. Vet. Rec. 126(23), 570–572.

Gregory, N. G. and Wotton, S. B. 1990b. Effect of stunning on spontaneous physical activity and evoked activity in the brain. Br. Poult. Sci. 31(1), 215–220. https://doi.org/10.1080/00071669008417248.

Gregory, N. G. and Wotton, S. B. 1991. Effect of a 350 Hz DC stunning current on evoked responses in the chicken's brain. *Res. Vet. Sci.* 50(2), 250–251. https://doi.org/10.1016/0034-5288(91)90118-8.

Gregory, N. G. and Wotton, S. B. 1994. Effect of electrical stunning current on the duration of insensibility in hens. *Br. Poult. Sci.* 35(3), 463–465. https://doi.org/10.1080/00071669408417711.

Grilli, C., Loschi, A. R., Rea, S., Stocchi, R., Leoni, L. and Conti, F. 2015. Welfare indicators during broiler slaughtering. *Br. Poult. Sci.* 56(1), 1–5. https://doi.org/10.1080/00071668.2014.991274.

Hillebrand, S. J., Lambooy, E. and Veerkamp, C. H. 1996. The effects of alternative electrical and mechanical stunning methods on hemorrhaging and meat quality of broiler breast and thigh muscles. *Poult. Sci.* 75(5), 664–671. https://doi.org/10.3382/ps.0750664.

Hindle, V. A., Lambooij, E., Reimert, H. G. M., Workel, L. D. and Gerritzen, M. A. 2010. Animal welfare concerns during the use of the water bath for stunning broilers, hens, and ducks. *Poult. Sci.* 89(3), 401–412. https://doi.org/10.3382/ps.2009-00297.

Holloway, P. H. and Pritchard, D. G. 2017. Effects of ambient temperature and water vapor on chamber pressure and oxygen level during low atmospheric pressure stunning of poultry. *Poult. Sci.* 96(8), 2528–2539. https://doi.org/10.3382/ps/pex066.

Hughes, B. O. 1983. Headshaking in fowls: the effect of environmental stimuli. *Appl. Anim. Ethol.* 11(1), 45–53. https://doi.org/10.1016/0304-3762(83)90078-0.

Humane Slaughter Association 2016a. Maintaining an uninterrupted electrical circuit and optimising current flow [WWW Document]. HSA online guide – electrical waterbath stunning poultry. Available at: https://www.hsa.org.uk/electrical-waterbath-stunning-of-poultry-introduction/introduction-7 (Accessed 9 March 2020).

Humane Slaughter Association 2016b. *Electrical Waterbath Stunning of Poultry*. Humane Slaughter Association, Wheathampstead, UK.

Humane Slaughter Association 2016c. *Practical Slaughter of Poultry*. Humane Slaughter Association, Wheathampstead, UK.

Hunter, R. R., Mitchell, M. A., Carlisle, A. J., Quinn, A. D., Kettlewell, P. J., Knowles, T. G. and Warriss, P. D. 1998. Physiological responses of broilers to pre-slaughter lairage: effects of the thermal micro-environment? *Br. Poult. Sci.* 39, 53–54. https://doi.org/10.1080/00071669888377.

Jacobs, L., Bourassa, D. V., Harris, C. E. and Buhr, R. J. 2019. Euthanasia: manual Versus mechanical cervical dislocation for broilers. *Animals* 9(2), 47. https://doi.org/10.3390/ani9020047.

Jacobs, L., Delezie, E., Duchateau, L., Goethals, K. and Tuyttens, F. A. M. 2017. Impact of the separate pre-slaughter stages on broiler chicken welfare. *Poult. Sci.* 96(2), 266–273. https://doi.org/10.3382/ps/pew361.

Jiang, N. N., Xing, T., Wang, P., Xie, C. and Xu, X. L. 2015. Effects of water-misting sprays with forced ventilation after transport during summer on meat quality, stress parameters, glycolytic potential and microstructures of muscle in broilers. *Asian-Australas. J. Anim. Sci.* 28(12), 1767–1773. https://doi.org/10.5713/ajas.15.0152.

Johnson, C. L. C. 2014. A review of bird welfare during controlled atmosphere and electrical water-bath stunning. *J. Am. Vet. Med. Assoc.* 245(1), 60–68. https://doi.org/10.2460/javma.245.1.60.

Kannan, G., Heath, J. L., Wabeck, C. J. and Mench, J. A. 1997. Shackling of broilers: effects on stress responses and breast meat quality. *Br. Poult. Sci.* 38(4), 323–332. https://doi.org/10.1080/00071669708417998.

Kettlewell, P. J. and Mitchell, M. A. 1994. Catching, handling and loading of poultry for road transportation. *World's Poult. Sci. J.* 50(1), 54–56. https://doi.org/10.1079/WPS19940005.

Kirkden, R., Niel, L., Stewart, S. and Weary, D. 2008. Gas killing of rats: the effect of supplemental oxygen on aversion to carbon dioxide. *Animal Welfare*, 17(1), 79–87.

Kittelsen, K. E., Granquist, E. G., Vasdal, G., Tolo, E. and Moe, R. O. 2015. Effects of catching and transportation versus pre-slaughter handling at the abattoir on the prevalence of wing fractures in broilers. *Anim. Welf.* 24(4), 387–389. https://doi.org/10.7120/09627286.24.4.387.

Knowles, T. G. and Broom, D. M. 1990. The handling and transport of broilers and spent hens. *Appl. Anim. Behav. Sci.* 28(1–2), 75–91. https://doi.org/10.1016/0168-1591(90)90047-H.

Knowles, T. G. and Wilkins, L. J. 1998. The problem of broken bones during the handling of laying hens – a review. *Poult. Sci.* 77(12), 1798–1802.

Korte, S. M., Beuving, G., Ruesink, W. and Blokhuis, H. J. 1997. Plasma catecholamine and corticosterone levels during manual restraint in chicks from a high and low feather pecking line of laying hens. *Physiol. Behav.* 62(3), 437–441. https://doi.org/10.1016/S0031-9384(97)00149-2.

Kranen, R. W., Veerkamp, C. H., Lambooy, E., Van Kuppevelt, T. H. and Veerkamp, J. H. 1996. Hemorrhages in muscles of broiler chickens: the relationships among blood variables at various rearing temperature regimens. *Poult. Sci.* 75(4), 570–576. https://doi.org/10.3382/PS.0750570.

Lambooij, B. and Hindle, V. 2018. Electrical stunning of poultry. In *Advances in Poultry Welfare*. Woodhead Publishing, UK, pp. 77–98. https://doi.org/10.1016/B978-0-08-100915-4.00004-X.

Lambooij, E., Anil, H., Butler, S. R., Reimert, H., Workel, L., Hindle, V. and Ur, W. 2011. Transcranial magnetic stunning of broilers: a preliminary trial to induce unconsciousness. *Animal Welfare*, 20(3), 407–412.

Lambooij, E., Reimert, H. G. M. and Hindle, V. A. 2010. Evaluation of head-only electrical stunning for practical application: assessment of neural and meat quality parameters. *Poult. Sci.* 89(12), 2551–2558. https://doi.org/10.3382/PS.2010-00815.

Lambooij, E., Reimert, H. G. M., Workel, L. D. and Hindle, V. A. 2012. Head-cloaca controlled current stunning: assessment of brain and heart activity and meat quality. *Br. Poult. Sci.* 53(2), 168–174. https://doi.org/10.1080/00071668.2012.665434.

Leary, S., Underwood, W., Anthony, R., Cartner, S., Grandin, T., Greenacre, C., Gwaltney-Brant, S., McCrackin, M. A., Meyer, R., Miller, D., Shearer, J., Turner, T. and Yanong, R. 2007. AVMA guidelines on euthanasia. *Am. Vet. Med. Assoc. Schaumburg* 218, 1–39. https://doi.org/10.1007/s11634-008-0026-3.

Leary, S., Underwood, W., Anthony, R., Cartner, S., Grandin, T., Greenacre, C., Gwaltney-Brant, S., McCrackin, M. A., Meyer, R., Miller, D., Shearer, J., Turner, T., Yanong, R., Johnson, C. L. and Patterson-Kane, E. 2020. *AVMA Guidelines for the Euthanasia of Animals: 2020 Edition*. AVMA American Veterinary Medical Association, Schaumburg, IL.

Lewis, P. D. and Morris, T. R. 2000. Poultry and coloured light. *Worlds. Poult. Sci. J.* 56(3), 189–207. https://doi.org/10.1079/WPS20000015.

Lines, J. A., Berry, P., Cook, P., Schofield, C. P. and Knowles, T. G. 2012. Improving the poultry shackle line. *Anim. Welf.* 21(1), 69–74. https://doi.org/10.7120/096272 812X13353700593608.

Lines, J. A., Jones, T. A., Berry, P. S., Cook, P., Spence, J. and Schofield, C. P. 2011a. Evaluation of a breast support conveyor to improve poultry welfare on the shackle line. *Vet. Rec.* 168(5), 129. https://doi.org/10.1136/vr.c5431.

Lines, J. A., Raj, A. B., Wotton, S. B., O'Callaghan, M. and Knowles, T. G. 2011b. Head-only electrical stunning of poultry using a waterbath: a feasibility study. *Br. Poult. Sci.* 52(4), 432–438. https://doi.org/10.1080/00071668.2011.587180.

Llonch, P., Dalmau, A., Rodríguez, P., Manteca, X. and Velarde, A. 2012. Aversion to nitrogen and carbon dioxide mixtures for stunning pigs. *Anim. Welf.* 21(1), 33–39. https://doi.org/10.7120/096272812799129475.

Mackie, N. and McKeegan, D. E. F. 2016. Behavioural responses of broiler chickens during low atmospheric pressure stunning. *Appl. Anim. Behav. Sci.* 174, 90–98. https://doi .org/10.1016/j.applanim.2015.11.001.

Martin, J. E., Christensen, K., Vizzier-Thaxton, Y. and McKeegan, D. E. 2016a. Effects of light on responses to low atmospheric pressure stunning in broilers. *Br. Poult. Sci.* 57(5), 585–600. https://doi.org/10.1080/00071668.2016.1201200.

Martin, J. E., Christensen, K., Vizzier-Thaxton, Y. and McKeegan, D. E. F. 2016b. Effects of analgesic intervention on behavioural responses to low atmospheric pressure stunning. *Appl. Anim. Behav. Sci.* 180, 157–165. https://doi.org/10.1016/j.applanim. 2016.05.007.

Martin, J. E., Christensen, K., Vizzier-Thaxton, Y., Mitchell, M. A. and McKeegan, D. E. F. 2016c. Behavioural, brain and cardiac responses to hypobaric hypoxia in broiler chickens. *Physiol. Behav.* 163, 25–36. https://doi.org/10.1016/j.physbeh.2016.04 .038.

Martin, J. E., McKeegan, D. E. F., Magee, D. L., Armour, N. and Pritchard, D. G. 2019a. Pathological consequences of low atmospheric pressure stunning in broiler chickens. *Animal* 14(1), 129–137. https://doi.org/10.1017/S1751731119001411.

Martin, J. E., McKeegan, D. E. F., Sparrey, J. and Sandilands, V. 2017. Evaluation of the potential killing performance of novel percussive and cervical dislocation tools in chicken cadavers. *Br. Poult. Sci.* 58(3), 216–223. https://doi.org/10.1080/00071668 .2017.1280724.

Martin, J., McKeegan, D., Sparrey, J. and Sandilands, V. 2016d. Comparison of novel mechanical cervical dislocation and a modified captive bolt for on-farm killing of poultry on behavioural reflex responses and anatomical pathology. *Anim. Welf.* 25(2), 227–241. https://doi.org/10.7120/09627286.25.2.227.

Martin, J. E., Sandercock, D. A., Sandilands, V., Sparrey, J., Baker, L., Sparks, N. H. C. and McKeegan, D. E. F. 2018a. Welfare risks of repeated application of on-farm killing methods for poultry. *Animals* 8(3). https://doi.org/10.3390/ani8030039.

Martin, J. E., Sandilands, V., Sparrey, J., Baker, L., Dixon, L. M. and McKeegan, D. E. F. 2019b. Welfare assessment of novel on-farm killing methods for poultry. *PLoS ONE* 14(2), e0212872. https://doi.org/10.1371/journal.pone.0212872.

Martin, J. E., Sandilands, V., Sparrey, J., Baker, L. and McKeegan, D. E. F. 2018b. On farm evaluation of a novel mechanical cervical dislocation device for poultry. *Animals* 8(1). https://doi.org/10.3390/ani8010010.

McCulloch, S. P. 2018. Brexit and animal protection: legal and political context and a framework to assess impacts on animal welfare. *Animals* 8(11), 213. https://doi.org/10.3390/ani8110213.

McKeegan, D. 2018. Mass depopulation. *Adv. Poult. Welf.*, 351–372. https://doi.org/10.1016/B978-0-08-100915-4.00017-8.

McKeegan, D., McIntyre, J., Demmers, T., Lowe, J., Wathes, C., Van Den Broek, P., Coenen, A. and Gentle, M. 2007. Physiological and behavioural responses of broilers to controlled atmosphere stunning: implications for welfare. *Anim. Welf.* 16, 409–426.

McKeegan, D. E., Reimert, H. G., Hindle, V. A., Boulcott, P., Sparrey, J. M., Wathes, C. M., Demmers, T. G. and Gerritzen, M. A. 2013a. Physiological and behavioral responses of poultry exposed to gas-filled high expansion foam. *Poult. Sci.* 92(5), 1145–1154. https://doi.org/10.3382/PS.2012-02587.

McKeegan, D. E., Sparks, N. H., Sandilands, V., Demmers, T. G., Boulcott, P. and Wathes, C. M. 2011. Physiological responses of laying hens during whole-house killing with carbon dioxide. *Br. Poult. Sci.* 52(6), 645–657. https://doi.org/10.1080/00071668.2011.640307.

McKeegan, D. E. F. 2004. Mechano-chemical nociceptors in the avian trigeminal mucosa. *Brain Res. Rev.* 46(2), 146–154. https://doi.org/10.1016/J.BRAINRESREV.2004.07.012.

McKeegan, D. E. F., McIntyre, J., Demmers, T. G. M., Wathes, C. M. and Jones, R. B. 2006. Behavioural responses of broiler chickens during acute exposure to gaseous stimulation. *Appl. Anim. Behav. Sci.* 99(3–4), 271–286. https://doi.org/10.1016/j.applanim.2005.11.002.

McKeegan, D. E. F., Sandercock, D. A. and Gerritzen, M. A. 2013b. Physiological responses to low atmospheric pressure stunning and the implications for welfare. *Poult. Sci.* 92(4), 858–868. https://doi.org/10.3382/ps.2012-02749.

Meyn 2020. Multistage CO2 stunning system [WWW Document]. Available at: https://www.meyn.com/products/live-bird-handling/multistage-co2-stunning-system (Accessed 2.12.2020).

Milsom, W. K., Abe, A. S., Andradeb, D. V. and Tattersall, G. J. 2004. Evolutionary trends in airway CO2/H+ chemoreception. *Respir. Physiol. Neurobiol.* 144(2–3), 191–202. https://doi.org/10.1016/J.RESP.2004.06.021.

Mitchell, M. A. and Kettlewell, P. J. 1994. Road transportation of broiler chickens: induction of physiological stress. *World's Poult. Sci. J.* 50(1), 57–59. https://doi.org/10.1079/WPS19940006.

Mitchell, M. A. and Kettlewell, P. J. 1998. Physiological stress and welfare of broiler chickens in transit: solutions not problems! *Poult. Sci.* 77(12), 1803–1814.

Mohamed, R. A., Eltholth, M. M. and El-Saidy, N. R. 2014. Rearing broiler chickens under monochromatic blue light improve performance and reduce fear and stress during pre-slaughter handling and transportation. *Biotechnol. Anim. Husb.* 30(3), 457–471. https://doi.org/10.2298/BAH1403457M.

Moosavi, S. H., Golestanian, E., Binks, A. P., Lansing, R. W., Brown, R. and Banzett, R. B. 2003. Hypoxic and hypercapnic drives to breathe generate equivalent levels of air hunger in humans. *J. Appl. Physiol.* 94(1), 141–154. https://doi.org/10.1152/japplphysiol.00594.2002.

More, S., Bicout, D., Bøtner, A., Butterworth, A., Calistri, P., Depner, K., Edwards, S., Garin-Bastuji, B., Good, M., Gortazar Schmidt, C., Miranda, M.A., Nielsen, S.S., Sihvonen, L.,

Spoolder, H., Willeberg, P., Raj, M., Thulke, H., Velarde, A., Vyssotski, A., Winckler, C., Cortiñas Abrahantes, J., Garcia, A., Muñoz Guajardo, I., Zancanaro, G. and Michel, V. 2017. Low atmospheric pressure system for stunning broiler chickens. *EFSA J.* 15(12), e05056. https://doi.org/10.2903/j.efsa.2017.5056.

National Chicken Council (NCC) 2011. *National Chicken Council Animal Welfare Guidelines*. NCC, Washington, DC.

Newcomb, L. 2010. Why use oxygen on everest? The physiological advantages of using supplementary oxygen on the summit day. *BJA Br. J. Anaesth.* 105. https://doi.org/10.1093/bja/el_6439.

Nicol, C., Caplen, G., Edgar, J., Richards, G. and Browne, W. 2011. Relationships between multiple welfare indicators measured in individual chickens across different time periods and environments. *Anim. Welf.* 20, 133-143.

Nielsen, S. S., Alvarez, J., Bicout, D. J., Calistri, P., Depner, K., Drewe, J. A., Garin-Bastuji, B., Gonzales Rojas, J. L., Gortázar Schmidt, C., Miranda Chueca, M. Á., Roberts, H. C., Sihvonen, L. H., Spoolder, H., Stahl, K., Velarde Calvo, A., Viltrop, A., Winckler, C., Candiani, D., Fabris, C., Van der Stede, Y. and Michel, V. 2019a. Slaughter of animals: poultry. *EFSA J.* 17(11), e05849. https://doi.org/10.2903/j.efsa.2019.5849.

Nielsen, S. S., Alvarez, J., Bicout, D. J., Calistri, P., Depner, K., Drewe, J. A., Garin-Bastuji, B., Gonzales Rojas, J. L., Gortázar Schmidt, C., Miranda Chueca, M. Á., Roberts, H. C., Sihvonen, L. H., Spoolder, H., Stahl, K., Velarde Calvo, A., Viltrop, A., Winckler, C., Candiani, D., Fabris, C., Van der Stede, Y. and Michel, V. 2019b. Killing for purposes other than slaughter: poultry. *EFSA J.* 17(11), e05850. https://doi.org/10.2903/j.efsa.2019.5850.

Nijdam, E., Arens, P., Lambooij, E., Decuypere, E. and Stegeman, J. A. 2004. Factors influencing bruises and mortality of broilers during catching, transport, and lairage. *Poult. Sci.* 83(9), 1610-1615. https://doi.org/10.1093/PS/83.9.1610.

Novoa, M., Vázquez, L., Lage, A., González-Torres, I., Pérez-García, L. F., Cobas, N. and Lorenzo, J. M. 2019. Water-bath stunning process in broiler chickens: effects of voltage and intensity. *Span. J. Agric. Res.* 17(2). https://doi.org/10.5424/sjar/2019172-14576.

Perez-Palacios, S. and Wotton, S. B. 2006. Effects of salinity and the use of shackle/leg sprays on the electrical conductivity of a commercial waterbath stunner for broilers. *Vet. Rec.* 158(19), 654-657. https://doi.org/10.1136/vr.158.19.654.

Petracci, M., Bianchi, M., Cavani, C., Gaspari, P. and Lavazza, A. 2006. Preslaughter mortality in broiler chickens, turkeys, and spent hens under commercial slaughtering. *Poult. Sci.* 85(9), 1660-1664. https://doi.org/10.1093/ps/85.9.1660.

Pierre, L. N., Emilie, B., Boissy, A., Boivin, X., Calandreau, L., Delon, N., Deputte, B., Desmoulin-Canselier, S., Dunier, M., Faivre, N., Giurfa, M., Guichet, J.-L. and Terlouw, C. 2018. Animal consciousness. *Anim. Welf.* 27, 87. https://doi.org/10.2903/sp.efsa.2017.EN-1196.

Poole, G. H. and Fletcher, D. L. 1998. Comparison of a modified atmosphere stunning-killing system to conventional electrical stunning and killing on selected broiler breast muscle rigor development and meat quality attributes. *Poult. Sci.* 77(2), 342-347. https://doi.org/10.1093/PS/77.2.342.

Powell, F. and Scheid, P. 1989. Physiology of gas exchange in the avian respiratory system. In: King, A. and McLelland, J. (Eds), *Form and Function in Birds*. Academic Press, London, UK, pp. 393-437.

Prayitno, D. S., Phillips, C. J. and Omed, H. 1997a. The effects of color of lighting on the behavior and production of meat chickens. *Poult. Sci.* 76(3), 452–457. https://doi.org/10.1093/ps/76.3.452.

Prayitno, D. S., Phillips, C. J. and Stokes, D. K. 1997b. The effects of color and intensity of light on behavior and leg disorders in broiler chickens. *Poult. Sci.* 76(12), 1674–1681. https://doi.org/10.1093/ps/76.12.1674.

Prinz, S., van Oijen, G., Ehinger, F., Bessei, W. and Coenen, A. 2010a. Effects of waterbath stunning on the electroencephalograms and physical reflexes of broilers using a pulsed direct current. *Poult. Sci.* 89(6), 1275–1284. https://doi.org/10.3382/ps.2009-00136.

Prinz, S., Van Oijen, G., Ehinger, F., Bessei, W. and Coenen, A. 2012. Electrical waterbath stunning: influence of different waveform and voltage settings on the induction of unconsciousness and death in male and female broiler chickens. *Poult. Sci.* 91(4), 998–1008. https://doi.org/10.3382/PS.2009-00137.

Prinz, S., van Oijen, G., Ehinger, F., Coenen, A. and Bessei, W. 2010b. Electroencephalograms and physical reflexes of broilers after electrical waterbath stunning using an alternating current. *Poult. Sci.* 89(6), 1265–1274. https://doi.org/10.3382/ps.2009-00135.

Quinn, A. D., Kettlewell, P. J., Mitchell, M. A. and Knowles, T. 1998. Air movement and the thermal microclimates observed in poultry lairages. *Br. Poult. Sci.* 39(4), 469–476. https://doi.org/10.1080/00071669888610.

Raj, M. 1998. Welfare during stunning and slaughter of poultry. *Poult. Sci.* 77(12), 1815–1819. https://doi.org/10.1093/ps/77.12.1815.

Raj, M. 2003. A critical appraisal of electrical stunning in chickens. *World's Poult. Sci. J.* 59(1), 89–98. https://doi.org/10.1079/WPS20030005.

Raj, M. 2006. Recent developments in stunning and slaughter of poultry. *World's Poult. Sci. J.* 62(3), 467–484. https://doi.org/10.1079/WPS2005109.

Raj, M. 2009. Stunning and slaughter. In: Duncan, I. and Hawkins, P. (Eds), *The Welfare of Domestic Fowl and Other Captive Birds*. Springer, New York, pp. 259–279. https://doi.org/10.1007/78-90-481-3650-6.

Raj, M., and Gregory, N. G. 1991. Preferential feeding behaviour of hens in different gaseous atmospheres. *Br. Poult. Sci.* 32(1), 57–65. https://doi.org/10.1080/00071669108417327.

Raj, M., and Gregory, N. G. 1990a. Investigation into the batch stunning/killing of chickens using carbon dioxide or argon-induced hypoxia. *Res. Vet. Sci.* 49(3), 364–366. https://doi.org/10.1016/0034-5288(90)90075-F.

Raj, M., and Gregory, N. G. 1990b. Effect of rate of induction of carbon dioxide anaesthesia on the time of onset of unconsciousness and convulsions. *Res. Vet. Sci.* 49(3), 360–363. https://doi.org/10.1016/0034-5288(90)90074-E.

Raj, M. and O'Callaghan, M. 2001. Evaluation of a pneumatically operated captive bolt for stunning/killing broiler chickens. *Br. Poult. Sci.* 42(3), 295–299. https://doi.org/10.1080/00071660120055232.

Raj, M. and O'Callaghan, M. 2004. Effect of amount and frequency of head-only stunning currents on the electroencephalogram and somatosensory evoked potentials in broilers. *Anim. Welf.* 13, 159–170.

Raj, M., O'Callaghan, M. and Hughes, S. 2006a. The effects of amount and frequency of pulsed direct current used in water bath stunning and of slaughter methods on spontaneous electroencephalograms in broilers. *Anim. Welf.* 15, 19-24.

Raj, M., O'Callaghan, M. and Hughes, S. 2006b. The effects of pulse width of a direct current used in water bath stunning and of slaughter methods on spontaneous electroencephalograms in broilers. *Anim. Welf.* 15, 25-30.

Raj, M., O'Callaghan, M. and Knowles, T. 2006c. The effects of amount and frequency of alternating current used in water bath stunning and of slaughter methods on electroencephalograms in broilers. *Anim. Welf.* 15, 7-18.

Raj, M., Sandilands, V. and Sparks, N. H. 2006d. Review of gaseous methods of killing poultry on-farm for disease control purposes. *Vet. Rec.* 159(8), 229-235. https://doi .org/10.1136/vr.159.8.229.

Raj, M. and Tserveni-Gousi, A. 2000. Stunning methods for poultry. *Worlds. Poult. Sci. J.* 56(4), 291-304. https://doi.org/10.1079/WPS20000021.

Raj, M., Wotton, S. B. and Gregory, N. G. 1992. Changes in the somatosensory evoked potentials and spontaneous electroencephalogram of hens during stunning with a carbon dioxide and argon mixture. *Br. Vet. J.* 148(2), 147-156. https://doi.org/10.1 016/0007-1935(92)90106-B.

Rao, A. 2014. Prestun shocks and mis-stuns during conventional slaughter. *Vet. Rec.* 174(18), 457-458. https://doi.org/10.1136/vr.g3016.

Rao, M., Knowles, T. and Wotton, S. 2013. The effect of pre-stun shocks in electrical water-bath stunners on carcase and meat quality in broilers. *Anim. Welf.* 22(1), 79-84. https ://doi.org/10.7120/09627286.22.1.079.

Rodrigues, D. R., Café, M. B., Jardim Filho, R. M., Oliveira, E., Trentin, T. C., Martins, D. B., Minafra, C. S. and Rodrigues, D. R. 2017. Metabolism of broilers subjected to different lairage times at the abattoir and its relationship with broiler meat quality. *Arq. Bras. Med. Vet. Zootec.* 69, 733-741. https://doi.org/10.1590/1678-4162-9268.

Rodríguez, P., Dalmau, A., Ruiz-De-La-Torre, J., Manteca, X., Litvan, H. and Velarde, A. 2008. Assessment of unconsciousness during carbon dioxide stunning in pigs. *Animal Welfare*, 17(4), 341-349.

Sandercock, D. A., Auckburally, A., Flaherty, D., Sandilands, V. and McKeegan, D. E. F. 2014. Avian reflex and electroencephalogram responses in different states of consciousness. *Physiol. Behav.* 133, 252-259. https://doi.org/10.1016/J.PHYSBEH.2 014.05.030.

Sandercock, D. A., Hunter, R. R., Nute, G. R., Mitchell, M. A. and Hocking, P. M. 2001. Acute heat stress-induced alterations in blood acid-base status and skeletal muscle membrane integrity in broiler chickens at two ages: implications for meat quality. *Poult. Sci.* 80(4), 418-425. https://doi.org/10.1093/PS/80.4.418.

Satterlee, D. G., Parker, L. H., Castille, S. A., Cadd, G. G. and Jones, R. B. 2000. Struggling behavior in shackled male and female broiler chickens. *Poult. Sci.* 79(5), 652-655. https://doi.org/10.1093/PS/79.5.652.

Schutt-Abraham, I., Wormuth, H. and Fessel, J. 1983. Electrical stunning of poultry in view of animal welfare and meat production. In: Eikelenboom, G. (Ed.), *Stunning of Animals for Slaughter*. Martinus Nijhoff, The Hague, The Netherlands, p. 154.

Shields, S. J. and Raj, M. 2010. A critical review of electrical water-bath stun systems for poultry slaughter and recent developments in alternative technologies. *J. Appl. Anim. Welf. Sci.* 13(4), 281-299. https://doi.org/10.1080/10888705.2010.507119.

Singer, P. 2016. *The Ethics of Killing Animals*. Oxford University Press, New York.

Small, A., Lea, J., Niemeyer, D., Hughes, J., McLean, D., McLean, J. and Ralph, J. 2019. Development of a microwave stunning system for cattle 2: Preliminary observations on behavioural responses and EEG. *Res. Vet. Sci.* 122, 72–80. https://doi.org/10.1016/J.RVSC.2018.11.010.

Small, A., McLean, D., Keates, H., Owen, J. S. and Ralph, J. 2013. Preliminary investigations into the use of microwave energy for reversible stunning of sheep. *Anim. Welf.* 22(2), 291–296. https://doi.org/10.7120/09627286.22.2.291.

Sparrey, J. M., Paice, M. E. R. and Kettlewell, P. J. 1992. Model of current pathways in electrical water bath stunners used for poultry. *Br. Poult. Sci.* 33(5), 907–916. https://doi.org/10.1080/00071669208417534.

Sparrey, J. M., Sandercock, D. A., Sparks, N. H. C. and Sandilands, V. 2014. Current and novel methods for killing poultry individually on-farm. *World's Poult. Sci. J.* 70(4), 737–758. https://doi.org/10.1017/S0043933914000816.

Sparrey, J. M. and Kettlewell, P. J. 1994. Shackling of poultry: is it a welfare problem? *Worlds. Poult. Sci. J.* 50(2), 167–176. https://doi.org/10.1079/WPS19940014.

Sparrey, J. M., Ketylewell, P. J., Paice, M. E. R. and Whetlor, W. C. 1993. Development of a constant current water bath stunner for poultry processing. *J. Agric. Eng. Res.* 56(4), 267–274. https://doi.org/10.1006/JAER.1993.1078.

Steiner, A. R., Flammer, S. A., Beausoleil, N. J., Berg, C., Bettschart-Wolfensberger, R., Pinillos, R. G., Golledge, H. D. W., Marahrens, M., Meyer, R., Schnitzer, T., Toscano, M. J., Turner, P. V., Weary, D. M. and Gent, T. C. 2019. Humanely ending the life of animals: Research priorities to identify alternatives to carbon dioxide. *Animals* 9(11), 1–25. https://doi.org/10.3390/ani9110911.

Terlouw, C., Bourguet, C. and Deiss, V. 2016. Consciousness, unconsciousness and death in the context of slaughter. Part I. Neurobiological mechanisms underlying stunning and killing. *Meat Sci.* 118, 133–146. https://doi.org/10.1016/J.MEATSCI.2016.03.011.

Tinker, D., Berry, P., Rycroft, J. and Sparks, N. 2004. Handling and catching of hens during depopulation. In: Perry, G. (Ed.), *Welfare of the Laying Hen*. CABI Publishing, Oxford, UK, pp. 345–360.

Tinker, D., Berry, P., White, R., Prescott, N., Welch, S. and Lankhaar, J. 2005. Improvement of the welfare of broilers by changes to a mechanical unloading system. *J. Appl. Poult. Res.* 14(2), 330–337 https://doi.org/10.1093/japr/14.2.330.

United States Department of Agriculture 1958. *Humane Methods of Livestock Slaughter Act*. USDA, Washington, DC.

United States Department of Agriculture 2019. *Poultry Slaughter: 2018 Summary*. USDA, Washington, DC.

US Department of Transportation: Federal Aviation Administration 2008. Aeromedical factors. *Pilot's Handbook of Aeronautical Knowledge* [WWW Document]. Available at: https://www.faa.gov/regulations_policies/handbooks_manuals/aviation/phak/ (Accessed 8 January 2020).

Velarde, A. and Dalmau, A. 2018. Slaughter without stunning. In: Mench, J. A. (Ed.), *Advances in Agricultural Animal Welfare*. Woodhead Publishing, Duxford, UK, pp. 221–240.

Velarde, A., Rodriguez, P., Dalmau, A., Fuentes, C., Llonch, P., von Holleben, K. V. V., Anil, M. H. H., Lambooij, J. B. B., Pleiter, H., Yesildere, T. and Cenci-Goga, B. T. T. 2014. Religious slaughter: evaluation of current practices in selected countries. *Meat Sci.* 96(1), 278–287. https://doi.org/10.1016/j.meatsci.2013.07.013.

Verhoeven, M. T. W., Gerritzen, M. A., Hellebrekers, L. J. and Kemp, B. 2014. Indicators used in livestock to assess unconsciousness after stunning: a review. *Animal* 9(2), 320–330. https://doi.org/10.1017/S1751731114002596.

Vieira, A., Vieira, F., Silva, I. and Filho, J. 2010. Productive losses on broiler preslaughter operations: effects of the distance from farms to abattoirs and of lairage time in a climatized holding area Perdas produtivas nas operações pré-abate de frangos de corte: efeito da distância entre as granjas e os. *Rev. Bras. Zootec.* 39, 2471.

Vieira, F. M., Silva, I. J., Barbosa Filho, J. A., Vieira, A. M. and Broom, D. M. 2011a. Preslaughter mortality of broilers in relation to lairage and season in a subtropical climate. *Poult. Sci.* 90(10), 2127–2133. https://doi.org/10.3382/PS.2010-01170.

Vieira, F. M. C., Silva, I. J. Od, Barbosa Filho, J. A. D., Vieira, A. M. C., Rodrigues-Sarnighausen, V. C. and Garcia, D. 2011b. Thermal stress related with mortality rates on broilers' preslaughter operations: a lairage time effect study. *Cienc. Rural* 41(9), 1639–1644. https://doi.org/10.1590/S0103-84782011005000111.

Villarroel, M., Pomares, F., Ibáñez, M. A., Lage, A., Martínez-Guijarro, P., Méndez, J. and de Blas, C. 2018. Rearing, bird type and pre-slaughter transport conditions. I. Effect on dead on arrival. *Span. J. Agric. Res.* 16(2). https://doi.org/10.5424/sjar/2018162-12013.

Vizzier-Thaxton, Y., Christensen, K. D., Schilling, M. W., Buhr, R. J. and Thaxton, J. P. 2010. A new humane method of stunning broilers using low atmospheric pressure. *J. Appl. Poult. Res.* 19(4), 341–348. https://doi.org/10.3382/japr.2010-00184.

Vogel, K. D., Badtram, G., Claus, J. R., Grandin, T., Turpin, S., Weyker, R. E. and Voogd, E. 2011. Head-only followed by cardiac arrest electrical stunning is an effective alternative to head-only electrical stunning in pigs. *J. Anim. Sci.* 89(5), 1412–1418. https://doi.org/10.2527/jas.2010-2920.

Warriss, P. D., Bevis, E. A., Brown, S. N. and Edwards, J. E. 1992. Longer journeys to processing plants are associated with higher mortality in broiler chickens. *Br. Poult. Sci.* 33(1), 201–206. https://doi.org/10.1080/00071669208417458.

Warriss, P. D., Knowles, T. G., Brown, S. N., Edwards, J. E., Kettlewell, P. J., Mitchell, M. A. and Baxter, C. A. 1999. Effects of lairage time on body temperature and glycogen reserves of broiler chickens held in transport modules. *Vet. Rec.* 145(8), 218–222. https://doi.org/10.1136/vr.145.8.218.

Warriss, P. D., Pagazaurtundua, A. and Brown, S. N. 2005. Relationship between maximum daily temperature and mortality of broiler chickens during transport and lairage. *Br. Poult. Sci.* 46(6), 647–651. https://doi.org/10.1080/00071660500393868.

Webster, A. B. and Fletcher, D. L. 2001. Reactions of laying hens and broilers to different gases used for stunning poultry. *Poult. Sci.* 80(9), 1371–1377. https://doi.org/10.1093/PS/80.9.1371.

Wideman, R. F., Rhoads, D. D., Erf, G. F. and Anthony, N. B. 2013. Pulmonary arterial hypertension (ascites syndrome) in broilers: a review. *Poult. Sci.* 92(1), 64–83. https://doi.org/10.3382/PS.2012-02745.

Wilkins, L. J., Wotton, S. B., Parkman, I. D., Kettlewell, P. J. and Griffiths, P. 1999. Constant current stunning effects on bird welfare and carcass quality. *J. Appl. Poult. Res.* 8(4), 465–471. https://doi.org/10.1093/japr/8.4.465.

Woolcott, C. R., Torrey, S., Turner, P. V., Serpa, L., Schwean-Lardner, K. and Widowski, T. M. 2018. Evaluation of two models of non-penetrating captive bolt devices for on-farm euthanasia of turkeys. *Animals* 8(3), 42. https://doi.org/10.3390/ani8030042.

Woolley, S. C., Borthwick, F. J. and Gentle, M. J. 1986a. Tissue resistivities and current pathways and their importance in pre-slaughter stunning of chickens. *Br. Poult. Sci.* 27(2), 301–306. https://doi.org/10.1080/00071668608416882.

Woolley, S. C., Borthwick, F. J. W. and Gentle, M. J. 1986b. Flow routes of electric currents in domestic hens during pre-slaughter stunning. *Br. Poult. Sci.* 27(3), 403–408. https://doi.org/10.1080/00071668608416896.

World Organisation for Animal Health (OIE) 2019. *Terrestrial Animal Health Code*. OIE, Paris. Available at: www.oie.int/en/international-standard-setting/terrestrial-code/access-online/ (Accessed on 4.1.2020).

Wotton, S., Zhang, X., McKinstry, J., Velarde, A. and Knowles, T. 2014. The effect of the required current/frequency combinations (EC 1099/2009) on the incidence of cardiac arrest in broilers stunned and slaughtered for the halal market. *PeerJ Prepr.* 2, e255v1. https://doi.org/10.7287/peerj.preprints.255v1.

Chapter 5

Humane transport, lairage and slaughter of sheep

P. H. Hemsworth and E. C. Jongman, University of Melbourne, Australia

1 Introduction

Contemporary public concerns about and policy debates on animal production focus on conditions that guarantee food security, public health, environmental quality and animal welfare (Vanhonacker et al., 2012). In general, societal concerns dictate the need for animal welfare standards and animal welfare legislation, but consumers and stakeholders in the meat supply value chain not only affect the supply and market shares of animal food products but also contribute to the growing disconnection between official (governmental or intergovernmental) welfare standards and those developed in the private sector, for example, by food industry stakeholders or NGOs (OIE, 2010; Vanhonacker et al., 2014).

http://dx.doi.org/10.19103/AS.2016.0019.19

Since most sheep in the world are grazed outdoors all year round, sheep farming is often perceived by the public and stakeholders as having few animal welfare problems relative to intensive animal production. The main reasons for this perception include the notion that outdoor production is natural, that it allows sheep to express their natural behaviours; and that sheep have evolved in these extensive conditions and are thus well adapted to this environment (Dwyer and Lawrence, 2008; Turner and Dwyer, 2007). Some farm animal housing and husbandry practices are contentious animal welfare issues for many, but there are also increasing community concerns about the treatment of farm animals post-farm gate, particularly animal transport and slaughter (de Jonge and van Trijp, 2013). For example, a survey across seven European countries indicated greater concern about methods of farm animal transport and slaughter than about farm animal husbandry (Kjaernes and Lavik, 2008). This chapter reviews the main welfare issues associated with management of sheep post-farm gate.

There is general agreement that some of the components involved in the transport and slaughter of sheep are the most controversial animal welfare issues post-farm gate (Grandin, 2007; Kjaernes and Lavik, 2008). Some of these issues have been shown to result in adverse consequences and thus reduced sheep welfare, whereas others are unresolved or may prove to be perceptions only. The issues that appear to be most frequently raised include floor space allowance, journey duration and thermal conditions for both road and sea transport; inanition and salmonellosis for long distance sea transport; and floor space allowance, properly trained stockpeople and proper restraint at slaughter with correct placement of the captive bolt or electrodes to ensure a more effective stun at abattoirs. The scientific basis or otherwise for these concerns will be considered here.

2 Animal welfare and its assessment

There are a number of reviews on animal welfare and its assessment (e.g., Boissy et al., 2007; Fraser, 2008; Mellor et al., 2009; Mellor, 2012; Hemsworth et al., 2015), and the following is an overview of this scientific perspective.

Animal welfare is a state and it is generally agreed that animal welfare relates to experienced sensations, that is, how the animal feels. These experiences are all subjective, varying in their affective or emotional contents and, based on human experience, are likely to include negative affective experiences such as thirst, hunger, nausea, pain and fear, and positive affective experiences such as satiety, contentment, companionship, curiosity and playfulness.

The conceptual framework of biological functioning has been predominantly used by scientists to assess animal welfare post-farm gate. The rationale for the approach is that difficult or inadequate adaptation will generate welfare problems for animals. This conceptual framework emphasises

Published by Burleigh Dodds Science Publishing Limited, 2021.

that animals use a range of behavioural and physiological responses to assist them to cope with challenges; while biological regulation in response to challenges occurs continuously, successful adaptation is not always possible. Marked challenges may overwhelm an individual's capacity to adapt and lead to its death. However, less severe challenges can still have significant biological costs, leading to impairment of growth, reproduction and health, which may reflect and/or result in welfare problems for the animal. Thus in environments in which adaptation is difficult, animal welfare is at risk.

The measures used to assess animal welfare post-farm gate include behavioural variables, such as fear, pain and illness behaviours; physiological variables, such as plasma concentrations of cortisol, neutrophil to lymphocyte ratios and immunoglobulin A; and fitness variables, such as lameness, skin lesions, disease and liveweight change. These measures have been used in this review to examine the main welfare issues associated with management of sheep post-farm gate.

3 Transporting sheep

3.1 Road transport

Road transport of livestock is largely a public activity, and is contentious for some. The effects of road transport on sheep welfare have been recently reviewed by Fisher et al. (2009), EFSA (2011) and Cockram (2014). These reviews and relevant research will be utilised here to highlight key welfare issues associated with transport. The main measures which have been used to assess the welfare implications arising from transport include mortality, weight loss, and physiological and behavioural responses. Key issues in relation to welfare during transport are handling during loading/unloading, floor space allowance, journey duration and environmental conditions.

Mortality of sheep during transport to a Queensland meatworks over an 11-year period ranged from 0.74 to 1.63% and mortality rate increased as annual rainfall decreased (Shorthose and Wythes, 1988). Mortality as a result of transport in less demanding environments is generally low, with rates of 0.006–0.018% being reported for sheep in the UK (Knowles, 1998). In the United Kingdom, mortality rate of lambs directly transported from farm to the abattoir was only 0.007% (Knowles et al., 1994); however, lambs arriving for slaughter from a livestock auction were over four times more likely to die in the abattoir lairage (housing), or to die during transport, than lambs which were sent directly from the farm. Changes in the mortality rate of the lambs from livestock auctions appeared to be associated with the price of slaughter lambs, and periods of increased mortality coincided with increased rates of carcase condemnations due to 'arthritis', 'abscess' and 'pleurisy'.

Weight loss during transport is generally attributed to deprivation of food and water, dehydration, excretion and tissue catabolism. Liveweight losses in sheep increase exponentially with time off feed and water. Typical liveweight losses in sheep range from 6 to 10% of body weight during road transport up to 24 hours (Broom et al., 1996; Knowles et al., 1998) and thus sheep appear to tolerate hunger and water deprivation extremely well. However, Hogan et al. (2007) suggest that further research is required since feed and water deprivation for 24 hours weakens the control of enteropathogenic bacteria by normal rumen microbes and thus may threaten animal health as well as meat safety. In an intensive study of stress physiology of sheep in environmental chambers, Parrott et al. (1996) found that deprivation of food and water for 48 hours over a wide ambient temperature range did not induce cortisol or prolactin release, while introduction to the chambers did so. Short-term deprivation of food and water is therefore unlikely to be the main source of stress during transport (Fisher et al., 2009).

Although sheep might show less obvious signs of distress during road transport than other species of farm animals, physiological responses to transport, such as plasma cortisol and adrenaline responses, are generally similar to other known psychological stressors such as isolation (Cockram, 2007) or to the effects of food and water deprivation alone (Knowles et al., 1995). However, loading and the early stages of transport increase heart rate, and plasma cortisol, prolactin and glucose concentrations in sheep, but after 3-9 hours are near to, or at, equivalent to pre-transport levels (Knowles et al., 1995; Broom et al., 1996; Parrott et al., 1998a), although these values will also be influenced by other factors such as the experience and condition of the animals, the driving events during the journey, and the duration of transport (EFSA, 2011).

Sheep show few behavioural responses to transport. They will ruminate, and if space permits, will lie down (Knowles, 1998). As stocking density increases, the proportion of sheep that lie down decreases (Cockram et al., 1996; Knowles et al., 1998). Sheep spend most of a 7-hour journey standing rather than lying down, but the amount of lying behaviour increases with journey duration (Cockram et al., 2004). On arrival after 12 hours of transport, sheep show increased eating, standing and drinking (Cockram et al., 1996). Similar responses were observed after 24 hours of transport, with frequencies of eating and drinking returning to normal after 14-16 hours and standing after 24 hours (Knowles et al., 1998). Cockram et al. (2000) found that sheep transported for 16 h from outside paddocks to inside pens had lower feed and water intakes than control sheep. However, there were no obvious effects on blood biochemistry, such as b-hydroxybutyrate, cortisol, total plasma protein, free fatty acids and osmolality, indicating that the novel lairage environment did not impair the ability of the sheep to recover from 16 h of transport.

Published by Burleigh Dodds Science Publishing Limited, 2021.

3.1.1 Floor space allowance

Space is one of the most important features of an animal's environment because it determines which behaviours the animals will be able to exhibit and for how long they exhibit them (Petherick, 2007). Some behaviours such as feeding, drinking and resting are obviously critical for immediate survival, whilst others such as locomotion/exercise, self-grooming and social behaviour are essential for longer-term health and welfare.

Petherick and Phillips (2009) used the research literature relating to transportation and intensive housing of sheep and cattle to assess the validity of allometric equations for estimating space allowances. In relation to transport, the authors proposed that for short-duration transportation during which animals remain standing, a space allowance per animal described by the allometric equation: area (m^2) = 0.020 $W^{0.66}$, where W= liveweight (kg), appears to be appropriate. Where it is desirable for all animals to lie simultaneously, they proposed a minimum space allowance per animal described by the allometric equation: area (m^2) = 0.027 $W^{0.66}$. Petherick and Phillips (2009) recommended that space allowances require verification with a range of species under different thermal conditions and, for transportation, under different conditions of vehicular/vessel stability.

Jones et al. (2010) examined the effects of space during transport of four categories of sheep, shorn and fleeced ewes and lambs. The behaviour of sheep in five space allowances was observed during standard journeys of 6 h: a low space allowance taken from the minimum space allowance in the European legislation (Anon., 1997); medium-low, medium-high and high allowances calculated from the allometric equation A = $kW^{0.67}$ (W: average liveweight/pen and k: empirical constant) with k-values of 0.021, 0.026 and 0.037, respectively; and a control group provided with more than 1m^2/animal. Fleeced sheep were given an additional 25% space. Sheep transported at control and high space allowances stood close to each other, but did not touch by bracing themselves against the motion of the vehicle by way of spreading their feet. Sheep fell to the floor mainly in low and medium-low space allowances, and more sheep were observed to voluntarily lie down at the control space allowance. The authors concluded that the space provided by minimum legislation and calculations with a k-value of 0.021 is unacceptable, as they do not allow sheep to adopt their preferred spacing strategy and lead to more losses of balance, slips and falls.

In a review of these findings, the European Food Safety Authority Panel on Animal Health and Welfare (EFSA, 2011) concluded that space allowances for sheep should be based on allometric equations relating size to body weight. Furthermore, it was recommended that the empirical coefficient (and space allowances) for journeys of 6 h with a mix of road types is: (i) shorn ewes, k =

0.026 (0.44 m² for 67 kg); (ii) fleeced ewes and lambs, $k = 0.033$ (0.56 m² for 65 kg, 0.4 m² for 40.5 kg); and (iii) shorn lambs, $k = 0.029$ (0.3 m² for 32.5 kg).

3.1.2 Journey duration

In a study of the physiological effects of 15 hours of transport, Broom et al. (1996) found that the major effects on plasma cortisol and prolactin concentrations occurred in the first 3 hours and the subsequent effects of transport were small. A more recent study by Fisher et al. (2010) examined the effects of 12, 30 and 48 hours of road transport of sheep on the physiological and behavioural responses as well as liveweight changes. While increasing duration of transport decreased liveweight, lack of effects of treatment on plasma cortisol concentrations and neutrophil to lymphocyte ratios by the end of the journey indicated that the physiological stress responses attributed to the early phases of transport were resolved by the end of the journeys. Furthermore, despite minimal lying during the journeys, sheep transported for 30 and 48 hours did not lie down more during the first 6 hours after arrival than sheep transported for 12 hours. The authors concluded that healthy adult sheep can tolerate transport and associated feed and water withdrawal periods of up to 48 hours without undue compromise to their welfare.

However, exposure to heat stress increases water loss principally through thermal panting, which increases the risk of significant dehydration (Thwaites, 1985); there has been considerable discussion around the value of rest stops. Short rest stops, such as one-hour rest is generally considered to provide little opportunity for recovery (Knowles et al., 1996; Hall et al., 1997; Jackson et al., 1999). Sheep may not drink water from unfamiliar sources in novel environments and sheep will eat first, before drinking; if feed is not available, drinking and resting are not priorities (Cockram et al., 1999a,b). A one-hour rest stop did not appear to serve its intended purpose, with many animals not drinking (Parrott et al., 1998b). Furthermore, rest stops may reduce animal fatigue and dehydration in the short term, but may add extra loading and unloading bouts and extend the overall duration of a journey (Fisher et al., 2009). Krawczel et al. (2007) concluded that rest stops involving unloading and provision of feed and water during long-distance transport (22 hours) at high ambient temperatures eliminated signs of food deprivation and maintained bodyweight but did not alleviate transport stress and evidence of immunosuppression. In addition, Knowles et al. (1995) concluded, based on liveweight and a range of stress and metabolic variables, that a single journey of 24 hours and, to a lesser extent, 34 hours was better for the welfare of sheep than an interrupted journey.

Cockram (2007) suggested that if care is taken only to select animals fit for transport, the environmental conditions (including driving style, road conditions, vehicle design and operation, space allowance, thermal conditions and ventilation), and the pre- and post-transport handling of the animals are optimal, it may be possible to transport animals over long distances without major welfare problems.

As Fisher et al. (2009) concludes, the question of whether a long journey is better managed by unloading the animals for rest stops with feed and water has no definitive answer for all situations. In some cases rest stops may address animal fatigue and hydration in the short term, but additional bouts of loading and unloading add to the cumulative stress and the overall duration of the journey.

3.1.3 Environmental conditions

A large part of the initial stress response to loading and transport may be the novelty of the procedure (Grandin, 1997). This has been supported by studies of other potential transport stressors, such as vehicular motion and noise, which induce similar effects in sheep (Hall et al., 1998). Evidence suggests that sheep show adaptive responses to transport within a few hours (Parrot et al., 1998a).

However, there is evidence that poor road conditions, such as rough and unpaved roads, may adversely affect sheep welfare by increasing heart rate (Ruiz-de-la-Torre et al., 2001) and plasma cortisol, lactate, glucose and creatine kinase concentrations (Miranda-de la Lama et al., 2011). Loss of balance and slipping occurred more on minor country roads than on larger country roads or motorways, particularly with low floor space allowances (Jones et al., 2010).

The effects of weather, and in particular heat stress, have generally been unexplored in hot climates, and could potentially have serious welfare implications. Models developed by Randall (1993) indicate that interactions between ambient temperature and humidity during transport are important at temperatures above 24°C, and that a temperature of 25°C should not be exceeded for unshorn sheep. These conditions would often be exceeded during livestock transport in many countries.

The temperature-humidity index (THI) inside the transport vehicle generally increases when vehicles are stationary in proportion to the duration of the stop. Fisher et al. (2005) reported that during journeys in summer in New Zealand, the stationary periods and the increase of external ambient temperature (>25°C) could induce thermal stress and be detrimental to the welfare of sheep. In this study, during stationary periods, 34% of THI readings exceeded 75, and, on average, the THI increased by 0.16 for every minute of a stationary period. Most trucks rely upon natural ventilation, which is generally proportional to the vehicle speed, but if the vehicle is stationary there may be no ventilation, and the thermal load imposed on the livestock will increase. Kettlewell et al. (2001) have demonstrated that installation of fans into a livestock truck resulted in adequate ventilation for the whole of the transit period. Since airflow is a critical factor in protecting sheep against heat stress, sheep transport vehicles should not be parked where airflow is absent or minimal, and duration of stops should obviously be minimised where possible.

In summary, transport is a stressful procedure, especially when combined with other procedures including mustering, yarding, handling, mixing and loading. The effects of transport on physiology such as heart rate; plasma cortisol, prolactin and glucose concentrations; and neutrophil to lymphocyte ratios have generally diminished after several hours, so that only the effects of food and water deprivation remain apparent. The key welfare issues for road transport are floor space allowance, journey duration and environmental conditions, such as THI, road conditions, and vehicle motion and noise. Hall and Bradshaw (1998) point out that transport may follow soon after other stressful experiences such as weaning, shearing, handling or marketing. Thus, interactions between transport and these procedures need evaluation.

3.2 Sea transport

Australia is the major exporter of ruminant livestock; other major regions exporting livestock by ship are the United States, southern South America, the Horn of Africa and Ireland (Phillips, 2008). During the 1980s about 6 million live sheep were annually exported from Australia to the Middle East (Kelly, 1995). The trade is still large, with over 2 million live sheep exported in 2014, with most (98%) exported to the Middle East (LiveCorp, 2014) on journeys lasting approximately 3 weeks.

The live export trade has been the subject of a great deal of academic and media interest for several decades. In a review of the ethical perspectives of the Australian live export trade, Phillips (2005) concluded that there are positive and negative aspects of the trade for each stakeholder group, and that the overall position adopted by any individual reflects their perspective of the balance of these components. Phillips also concluded that, while it seems likely that the trade is being scrutinised by many parties, not all parties have adequate information on the impact of the trade on animal welfare and other ethical considerations. Further, it was recommended that there is a need for openness on the part of the trade, a willingness to conduct research to improve the welfare of exported animals, and for unbiased, informative reporting on the part of the media.

As with land transport, key issues in relation to welfare during transport are handling during loading/unloading, space allowance, journey duration and environmental conditions. However, there is increased risk of cumulative stress during the extended period of transport. Furthermore, as Phillips and Santurtun (2013) recognise, the ship journey is a part of a much longer transport process. This usually includes mustering the animals, usually from grasslands, holding them before loading onto a vehicle, transfer to an assembly depot near the port, holding for several days in a feedlot to adapt to pelleted ration and high stocking densities before loading onto another vehicle to transfer to the port, offloading to enter the ship, the ship journey itself, offloading from the ship and loading onto another vehicle, and finally travel to a feedlot, where they remain for at least a few weeks before transport to an abattoir (Phillips, 2005). Sea transport is only part of the export process, and stressors both during and before and after sea transport contribute to the cumulative stress of the transport process, highlighted by mortalities.

The welfare implications of livestock transported by sea have been recently reviewed (Phillips and Santurtun, 2013; Foster and Overall, 2014) and these reviews and relevant research will be utilised here to highlight key welfare issues associated with sea transport. The measures used to assess animal welfare have included behavioural variables, such as open-mouthed panting, respiration rate and feeding behaviour; physiological variables, such as body temperature, plasma cortisol and haptoglobin concentrations, blood packed cell volume, red blood cells and platelet counts; and fitness variables, such as liveweight change. Furthermore, because of the long duration of transport, variables such as morbidity and mortality have been used in assessing animal welfare. While less severe challenges can still have significant biological costs, leading to liveweight loss, ill health and other impairments, which may reflect and/or result in welfare problems for the animal, marked challenges to homeostasis may overwhelm an individual's capacity to adapt and consequently lead to its death.

3.2.1 Inappetence, inanition, salmonellosis and mortality

The main welfare concern with long-distance sea transport has been mortality. Death rates for Australian sheep transported to Asia and the Middle East in the 1980s and 1990s were about 2% (Norris and Richards, 1989; Kelly, 1990, 1995; Norris et al., 1990b; Higgs et al., 1999) but mortality rates have been markedly reduced in the last decade. According to the Australian Department of Agriculture and Water Resources in reports to the Australian government, the average yearly mortality rate on all shipments of sheep was 0.78% (for 20.5 million sheep exported in 2009–15, Australian Government, 2016). This mortality rate at sea on a daily basis appears to be not too dissimilar from that of road transport in demanding conditions (see Land transport section).

The main cause of death with long-distance sea transport is starvation, which is variously described in the literature as shy feeding (Syme, 1985), failure to eat (Kelly, 1990), inanition or exhausted state due to prolonged undernutrition (Richards et al., 1989), and inappetence or loss of appetite (Norris et al., 1990c). Inanition predisposes sheep to death from salmonellosis (Higgs et al., 1993) and inanition and salmonellosis cause about 75% of all deaths at sea (Richards et al., 1989; Kelly, 1990; Norris et al., 1990c), while more recent evidence indicates that inanition and salmonellosis account for 43% and 20%, respectively, of deaths (Norris, 2005). Heat stress, in particular in Middle Eastern waters during the northern hemisphere summer, has been identified as another cause of mortality (Kelly, 1995; Caulfield et al., 2014).

Some of the risk factors associated with high mortality in Australian shipments to the Middle East have been identified. Risk increases with increasing voyage duration (Norris and Richards, 1989), increasing age and fatness of the animals (Higgs et al., 1991), and with decreasing acceptability of the pelleted ration (Norris et al., 1990b). Risk also varies with seasonal metabolic factors, with greater death rates in the second half of the year (Australian winter and spring, Higgs et al., 1991). Sheep coming from dry pasture in late autumn were metabolically adjusted to mobilise fat reserves. However, sheep coming from green pasture in late winter were gaining weight and were not adjusted to fat mobilisation (Richards et al., 1991). Sheep from farms in higher rainfall zones and with longer pasture-growing seasons were at greater risk (Higgs et al., 1999). In some voyages there was evidence of a 'tier effect', with sheep in the upper tier on every deck suffering higher mortalities than those in the lower tier (Kelly, 1995). High ambient temperatures, similar to those experienced by sheep travelling from Australia to the Middle East, were reported to reduce feed intake in some sheep (Stockman et al., 2011), but not as much as was observed with cattle in similar conditions (Phillips and Santurtun, 2013). Black (1996) reported that feed access, rather than inappetence in the presence of feed, limited feed intake in sheep travelling from New Zealand to the Middle East.

In summary, between 10 and 100% of sheep will refuse to eat a supplement they have never seen, and the intake of those that do eat is extremely variable (Lynch, 1988). Indeed, it is estimated that 5–20% of all sheep that enter a pre-embarkation feedlot do not adapt to the feed and the environment (Kirby et al., 2004; Jolly and Wallace, 2007). Furthermore, while the incidence of inappetence may be influenced by dominance behaviour, it may be influenced more by neophobia of either the trough or the feed itself and there is evidence of this in live export research (Norris et al., 1990a; Hodge et al., 1991). This cautious sampling or complete rejection of new foods is known as neophobia and is a well-known and well-reported phenomenon in ruminants. The number of sheep not feeding was not reduced by increasing the duration of lot-feeding (Norris et al., 1992) or by reducing competition for feed or trough space (Norris et al., 1990a). Neophobia can be manipulated using several techniques, including flavour generalisation and repeated exposure (Launchbaugh et al., 1997) and transmission of learned behaviour from ewe to lamb (Lynch and Bell, 1987). The effects of ship motion on sheep welfare are poorly understood, and Phillips (2008) suggests that motion sickness may contribute to inappetence.

Published by Burleigh Dodds Science Publishing Limited, 2021.

3.2.2 Floor space allowance

There is a lack of scientific evidence on the effects of space allowance during sea transport, but a starting point is to determine this from allometric relationships between bodyweight, space availability and behaviour (Petherick and Phillips, 2009) as has been done for land transport (see Land transport section). Indeed, Petherick and Phillips (2009) recommended that for intensive housing systems in which animals can lie simultaneously and move around and access food and water with ease, a minimum space allowance per animal described by the allometric equation, area (m²) = $0.033 W^{0.66}$, where W= liveweight (kg), appears to be the threshold below which there are adverse effects on welfare. However, Petherick and Phillips (2009) concluded that there are insufficient data to determine whether this allowance onboard a vessel would enable animals to move and access food and water with ease.

In two sea voyages from Australia to the Middle East, Ferguson and Lea (2013) examined the effects of three space allowances, the Australian Standards for the Export of Livestock (ASEL), AESL – 10% and the AESL+10% or the allometric allowance (0.027 (liveweight$^{0.66}$), whichever was greater. Space allowance was manipulated by varying group size. The mean liveweight and the floor space allowances for the three treatments were 47 ± 1.5 kg and 0.27, 0.31 and 0.34 m²/sheep, respectively, in the first voyage and 43 ± 1.5 kg and 0.27, 0.30 and 0.32 m²/sheep, respectively, in the second voyage. The overall levels of ill health and mortality was low (mortality less than 0.5%) on both voyages. There were no treatment effects on weight gain in either voyage. While there were no treatment effects on lying time in the second voyage, there was an interaction between treatment and day of voyage on lying time in the first voyage, with sheep in the low space allowance treatment (ASEL – 10%) spending less time lying than in the higher space allowance treatments during the initial stages of the voyage. The authors concluded that the current ASEL space allowances are appropriate but in the interests of continued refinement of live export standards, the putative benefits of a small increase above the ASEL space allowance, particularly during the critical early stages of a voyage, are worthy of further consideration and evaluation. There is evidence that feeder space affects both feeding behaviour and weight gain of sheep early in feedlots (Jongman et al., 2016). Since group size and thus feed and water trough space varied between treatments in the experiment by Ferguson and Lea (2013), research on feed and water trough space during sea transport is also warranted.

In general there is very little published evidence on the effects of space allowance in housed sheep. Gonyou et al. (1985) observed a 10% reduction in average daily gain in lambs weighing 27.7-28.8 kg when their space allowance was reduced from 0.48 m²/sheep to 0.32 m²/sheep over an 8-week week finishing period. The experiment by Ferguson and Lea (2013) was conducted during periods when climatic conditions were relatively benign, and thus further research particularly under different thermal conditions and different conditions of sea transport on behaviour and stress physiology would be judicious particularly considering the contentious nature of this practice.

3.2.3 High ammonia concentrations

In two ship voyages in which sheep were transported from Australia to the Middle East, Pines and Phillips (2011) found that high ammonia concentrations (up to 59 ppm) occur in those parts of the ship where there is insufficient ventilation and/or high ambient temperatures and humidity. It has been shown that exposure of sheep to ammonia concentrations of up to 15-45 ppm in simulated ship transport reduced feed intake and liveweight and increased pulmonary macrophage activity, compared to a control group (<6 ppm) (Phillips et al., 2012a). Furthermore, most sheep avoided concentrations of 45 ppm ammonia, with no evidence of habituation following a simulated 9-day voyage (Phillips et al., 2012b).

In summary, as with land transport, sea transport is stressful, which involves a number of procedures in addition to transport *per se* such as handling during loading/unloading, loading density, and environmental conditions. Furthermore, because of the extended period of transport there is increased risk of cumulative stress. The key welfare issues are inanition and salmonellosis, space allowance and environmental conditions. A contentious point for some in relation to live export is that it is difficult to ensure animal welfare in destination country because exported animals are beyond the export country's sovereign control (Fisher, 2013). Research on space allowances under different thermal conditions and different conditions of sea transport is required. While space may affect stress, access to feeders, inappetence and inanition, effects of ship motion on inappetence and welfare also require research. Furthermore, the effects of early experience of sheep on their adaptation to feed in the pre-embarkation feedlot and during transport as well as sea transport *per se* require research. Early identification of inappetence may further reduce the welfare risks during sea transport.

4 Abattoirs

There appears to be increasing community concern about the welfare of farm animals at abattoirs (Kjaernes and Lavik, 2008; de Jonge and van Trijp, 2013). The impetus to reduce stress prior to slaughter is that stress not only affects the welfare of sheep, but can also affect meat quality (Deiss et al., 2009; Ferguson et al., 2014).

Several factors may cause stress to animals at abattoirs pre-slaughter, such as the environment, management, facility design and the behaviour of stockpeople. Studies have shown that pre-slaughter handling may affect behaviour and stress of sheep and cattle (Hemsworth et al., 2011). There is a link between handling behaviour of stockpeople and their attitudes to handling and working with sheep and cattle at abattoirs (Coleman et al., 2012). Facility design may also influence animal handling, with poor design encouraging poor handling practices. Dogs are often used in abattoirs to move sheep during unloading and moving sheep from lairage all the way to the raceway leading to stunning. However, dogs can be a major stressor to sheep at abattoirs (Kilgour and de Langen, 1970; Hemsworth et al., 2011). In addition, stunning practices may be sub-optimal, due to method of stunning, variability in placement of electrodes, the design of the facility and the movement of animals at the time of attempted stunning.

While there is evidence that pre-slaughter stress, both acute and chronic, affect animal welfare and meat quality, little is known of the relative importance of the different aspects that can affect sheep welfare pre-slaughter. The next three sections review current knowledge on sheep handling and some associated factors that may affect sheep welfare at abattoirs.

5 Sheep pen design: rest and recovery in lairage

Lairage housing at the abattoir prior to slaughter provides the opportunity for rest and recovery from the effects of transport (Cockram et al., 1997). The key lairage requirements include sufficient space to lie down, sufficient time to allow recovery, and access to water to recover from dehydration. Lairage times can vary from several hours to more than 24 hours, depending on time of arrival, holding capacity and the number of animals required for slaughter. For example, in Europe and North America, sheep are generally slaughtered on the day of arrival at the abattoir, while in Australia sheep are often slaughtered the day after arrival or can be held for longer periods (Ferguson and Warner, 2008). When conditions in lairage are not optimal, dehydration and depletion of muscle glycogen can result, both of which are related to reduced meat quality (Toohey and Hopkins, 2006; Jacob et al., 2005, 2006).

The Australian Model Code of Practice for the Welfare of Animals: Livestock at Slaughtering Establishments (Anon., 2002) recommends that holding pens should provide no less than 0.6 m^2 per sheep. However, Weeks (2008) reported a space allowance in the United Kingdom as low as 0.23 m^2 per sheep. Reducing space allowance decreases lying behaviour and an allowance greater than 1 m^2 per sheep is required before all or most sheep in a pen lie down (Jarvis and Cockram, 1995; Kim et al., 1994; Jongman et al., 2008). Sheep do not distribute evenly throughout a pen and prefer to lie next to an open sided fence, rather than in the middle of the pen (Hargreaves and Hutson, 1997). Therefore pen shape may also influence lying behaviour.

Jarvis and Cockram (1995) observed the resting behaviour of sheep in lairage after four hours of transport and found a large variation in the number of sheep lying down. While the median was only 17%, lying behaviour ranged from 1 to 63% of sheep across different pens and over several hours. There was no effect of slats or straw on the total percentage of time spent lying down. However, the presence of human activity in the lairage area has been associated with increased alertness and movement and decreased lying behaviour (Kim et al., 1994). Similarly, Eldridge et al. (1989) recommended that movement of cattle in and out of lairage pens past resting animals should be minimised in order to minimise stress, and the same is likely to be true for sheep.

Very little is known about drinking behaviour of sheep in lairage, and sheep may be left for considerable time without water prior to and during transport. In addition, it has been suggested that the stress around transport and lairage, through elevated plasma cortisol concentrations, may induce higher urine output and prevent thirst in dehydrated animals (Hogan et al., 2007). Rehydration may therefore be limited during lairage. Thus water should be available in lairage and water troughs should be of sufficient size so that all animals can access drinking water within one hour of arrival (Anon., 2002). For example in a study at two different abattoirs in Australia, Jacob et al. (2006) found that up to 50% of lambs were dehydrated at slaughter, indicating that they failed to drink enough water to rehydrate following water deprivation during the farm curfew and transport periods. In addition, Jongman et al. (2008) found that 20% of sheep did not drink the available water during 24 hours in lairage, despite water deprivation of 15-28 hours and hot summer conditions. While space allowance ranging from 0.3 to 1 m^2/sheep did not affect access to water in the study, there is no information on the possible effect of trough size and placement. It is also unknown if the equivalent phenomenon of a 'shy feeder' (Savage et al., 2008) occurs in relation to drinking, that is, a 'shy drinker'. It is likely that the combination of stress, unfamiliarity of the water trough and change in taste of water (due to different water sources) may inhibit some sheep from drinking.

6 Sheep handling

6.1 Stockperson characteristics

Since the 1980s there has been an ever-increasing body of evidence on the profound effects of human interactions on farm animal fear and thus animal stress, productivity and welfare. This body of research has been recently reviewed by Hemsworth and Coleman (2011) and while most of the research has been conducted on farm animals in farm settings, this research has implications for handling and management of animals post-farm gate. Indeed limited research has shown that the attitudes and handling behaviour of stockpeople at abattoirs are related to sheep behaviour and stress pre-slaughter.

Recent research indicates that, while there is considerable variation, between- and within-abattoirs, pre-slaughter handling and sheep behaviour are associated with stress in sheep (Hemsworth et al., 2011, 2016). For example, plasma cortisol concentrations post-slaughter in sheep were best predicted by a mixture of stockperson and dog variables as well as lamb variables pre-slaughter. These relationships indicate that increased cortisol concentrations were associated with increased times that lambs spent from the forcing pen to the stunning area, in close visual contact with the stockperson, falling down and with the dog riding them. The identification of these predictor variables of cortisol, which may be a mixture of independent and mediating variables, supports the well-demonstrated effect of handling on fear and stress responses in livestock. These relationships, although not conclusive evidence of causal relationships, highlight the value of training stockpeople to reduce fear and stress in sheep at abattoirs.

Furthermore these relationships in these studies by Hemsworth et al. (2011, 2016) highlight that high levels of contact with stockpeople and dogs may be highly fear-provoking to sheep and that some sheep behaviours may be useful indicators of handling stress in sheep at abattoirs. Stockpeople in many countries have traditionally used dogs as a handling aid in moving sheep since dogs are likely to be perceived as a natural predator by sheep. It has been shown that of the stressors that sheep are commonly exposed to, dogs can cause the highest avoidance and cortisol response in sheep (Kilgour and de Langen, 1970; Beausoleil et al., 2005). Many of the close interactions with humans that sheep receive on farms are aversive, such as vaccinations, castration, tail docking and shearing; consequently it is not surprising, particularly for sheep raised in extensive systems, that close human contact is perceived as fear-provoking.

Understanding stressful handling practices, including use of dogs by abattoir stockpeople and their underlying aetiology is obviously critical in reducing handling risks to animal welfare, as well as meat quality. This will be considered later when discussing strategies post-farm gate to safeguard sheep welfare. Understanding the behavioural characteristics of sheep is also important for stockpeople in facilitating ease of movement and reducing handling stress in their sheep; this is highlighted in the next section.

Published by Burleigh Dodds Science Publishing Limited, 2021.

6.2 Sheep behavioural characteristics and ease of handling

Fear-like other affective states is affected by both internal and external stimuli. In relation to external stimuli, Gray (1987) recognizes that fear may be triggered by environmental stimuli which are novel; have high intensity such as loud and large stimuli; have special evolutionary dangers such as heights, isolation and darkness; and arise from social interaction such as contagious learning or have been paired with aversive experiences. While fear appears to be activated almost exclusively by external stimuli (Hogan, 2008), intrinsic factors are also important. For example, social, gender, breed and strain effects on fear responses have been shown in both laboratory and farm animal species (Hemsworth and Coleman, 2011).

Evidence from handling studies and observations on human–animal interactions in the livestock industries indicate that the history of interactions between humans and the animal leads to the development of a stimulus-specific response of farm animals to humans: through conditioning, farm animals may associate humans with rewarding and punishing events that occur at the time of human-animal interactions and thus conditioned responses to humans develop (Hemsworth et al., 2011). While it may intuitively appear that fearful animals may be easy to move, animals that are fearful of humans are generally the most difficult to handle particularly in unfamiliar locations. Animals that are fearful of both the handler and the unfamiliar location in which they are being moved are likely to show panic behaviours, such as baulking, fleeing and packing, making handling more difficult (Hemsworth and Coleman, 2011). While fear of humans may affect ease of handling, other factors such as the animal's familiarity and experience with the location and the handling behaviour of the stockperson at the time of handling, as well as the fear-provoking features in the environment in which they are being moved to, are also likely to affect ease of handling.

The flight distance of sheep to humans – that is, the distance at which sheep avoid an approaching human – depends not only on the fear of humans that sheep experience but also on the space available for escape. For example Hutson (1982) found that the flight distance in a 4-m-wide laneway was twice that compared to a 2-m-wide laneway.

Sheep can be moved in large groups due to their natural following behaviour. Because flocking and following is such a key feature of sheep behaviour, any handling that involves separating or disrupting groups of sheep may cause difficulty in handling. It is considered that a group of sheep should consist of at least five animals to express normal flock behaviour (Hargreaves and Hutson, 1997).

Sheep have very good eyesight and thus vision is an important factor when designing sheep-handling facilities (Hargreaves and Hutson, 1997). It is important to consider changes in light, visual cliffs and different colours

(Hargreaves and Hutson, 1997). While auditory cues are less important than visual cues and sheep generally habituate to constant noise, they do respond to intermittent or sudden noise. In addition, breed, sex and age all have an influence on ease of handling. For example, Njisane and Muchenje (2013) reported that Merinos and Merino crosses were calmer when handled than Dorpers, and ewes were calmer than castrates.

7 Stunning

Legislation in most countries require animals to be unconscious and insensible to pain so that slaughter can be performed without avoidable fear, anxiety, pain, suffering or distress (EFSA, 2006). Sheep are generally stunned using a penetrating captive bolt or electric stunning, with the latter being preferable, particularly with polled sheep (Terlouw et al., 2008). Electrical stunning will cause unconsciousness or death by electrocution, depending on placement of the electrodes and current applied, and electrical stunning is normally followed by neck incision to kill the animal. Effective restraint of animals during stunning can be one of the most stressful phases during the slaughter process (Rushen, 1986), but is essential to ensure the proper application of the mechanical or electrical stunning method. While electrical stunning can be carried out on unrestrained animals held in a group, this may reduce the effectiveness of the stun due to incorrect placement of the electrodes (Terlouw et al., 2008). For example, Berg et al. (2012) found that 59% of lambs in one trial and 26% in another were stunned using incorrect positioning of the tongs using head-only stunning with scissor-type stunning tongs.

While head-only electrical stunning with electrodes placed across the brain is the most common form of electrical stunning, an alternative method is head to back stunning, in which an electrical current is simultaneously passed through the brain and the heart. The latter method causes cardiac fibrillation and therefore death. Both stunning methods may be effective; however, head-to-back stunning has been found to be preferable by Gregory and Wotton (1984) as the neck incision following head-only electrical stunning may not always be effective, in terms of the interval from stun to both the neck incision and the severing of the two carotid arteries (Blackmore and Petersen, 1981; Gregory and Wotton, 1984).

There is an increase in the world market for halal meat and an increase in export of halal meat from non-Muslim countries. Head-only stunning is the only approved electrical stunning method for halal slaughter (Farouk et al., 2014; Nakyinsige et al., 2013) and is therefore a suitable method of stunning for use in export abattoirs in non-Muslim countries. In most sheep abattoirs the head-only stun is used by applying a handheld unit with prong electrodes or scissor-type stunning tongs, and animals are held either in small group pens or in a V-type restrainer while being stunned.

Signs of efficient stunning in sheep include tonic and clonic activity and absence of normal rhythmic breathing (Velarde et al., 2002). With head-only stunning, positioning the electrodes anywhere other than between the eyes and the base of the ears means that more of the current may flow through lower resistance pathways and not entirely through the brain, thus reducing the effectiveness of the stun (Anil, 2012). While the presence of wool, a dry skin, or placement of the tongs in caudal position behind the ears can affect the effectiveness of stunning (Velarde et al., 2000), less is known about prong electrodes. While tong electrodes were found to be ineffective on lambs with dry skin and wool (Velarde et al., 2000), prong electrodes penetrate wool more easily, and so may be more effective in this situation.

With small areas of contact between the sheep's head and the electrodes, wool-burning and marked carbonising of the electrodes can occur (Anil, 2012). This, in turn, leads to a poor electrical contact due to an increased electrical resistance in the pathway, and special care is necessary to keep the electrodes clean. Effective head-only stunning in sheep should be induced using a minimum current of 1.0 ampere, a minimum of 250 volts and a duration of current flow of a minimum of two seconds, and a maximum stun-to-neck incision interval of between 8 (EFSA, 2004) and 15 seconds (Anil and McKinstry, 1991).

7.1 Religious slaughter without stunning

In most western countries where stunning before slaughter is a legal requirement, there is debate about the exemption on religious grounds for shechita and certain halal certifications. There is a consensus of opinion in the scientific literature that a neck incision without stunning is very likely to be painful (Gregory et al., 2010; Johnson et al., 2012; Nakyinsige et al., 2013; Johnson et al., 2015). For example Gibson et al. (2009a) showed that a ventral-neck incision in calves was associated with evidence of significant noxious sensory input, as measured by changes in electroencephalographic responses and that this was associated with noxious stimuli (Gibson et al., 2009b). A neck incision does not result in instantaneous insensibility and sheep are conscious for at least 2–8 seconds after transection of the major arteries in the neck, but may be conscious for as long as 8–20 seconds (Johnson et al., 2015). Therefore slaughter without stunning poses a risk to animal welfare for this period of time.

In summary, from the time of unloading at the abattoir to the time of slaughter, sheep experience a number of stressors. Handling by stockpeople often involves negative interactions to facilitate movement and may involve the use of dogs. Well-designed facilities and properly trained stockpeople using judicious use of contact with sheep as well as judicious use of dogs can minimise negative interactions when handling, and therefore minimise stress in sheep prior to slaughter. Disturbance and overcrowding during lairage should be avoided and sheep should have access to water at all times while in lairage. While restraint during stunning may be stressful in itself, proper restraint assists with the correct placement of the captive bolt or electrodes, ensuring a more effective stun and an animal that is insensible during neck incision.

8 Safeguarding animal welfare

8.1 Science and education

Science provides the means to understand the impact of animal use on the animal and thus it has an important role in underpinning societal decisions on animal use and the acceptability or otherwise of attendant conditions and compromises. Obviously, societal interests in animal welfare also include consideration of wider issues such as human health, economic and social implications, as well as environmental impacts (Fisher and Mellor, 2008; Mellor and Bayvel, 2008). Decisions on acceptable animal use can therefore involve difficult and complex choices and consequently may remain controversial.

The public is often a key driver of animal welfare change since public views affect decision makers at the political, regulatory, retail and industry levels. However, public attitudes about animal welfare are often based on limited knowledge, and the public's beliefs are largely acquired from the mass media, perhaps filtered by opinion leaders (Coleman et al., 2015). Coleman (2010) concluded that to address both the mismatch between the public's perceived and actual knowledge of livestock practices and public welfare concerns, there needs to be accurate and reciprocal communication between the livestock industries and the community.

While science and community values underpin the establishment, amendment or validation of industry welfare standards and practices, it is critical to deliver industry education, through staff training. As discussed in the next section, training of stockpeople is essential in safeguarding animal welfare.

8.2 Training stockpeople

Stockpeople require a range of well-developed husbandry skills and knowledge to effectively care for and handle farm animals. Appreciating the factors that affect work performance, as well as where deficiencies exist is the first step in developing a strategic programme to ensure that stockpeople have well-developed husbandry skills and knowledge, as well as access to the appropriate facilities and opportunity to use these skills and knowledge to effectively care for farm animals (Hemsworth and Coleman, 2009, 2011).

Knowing and being skilled at the techniques that must be used to accomplish a task are clearly prerequisites to being able to perform that task and thus these job-related characteristics will be limiting factors on job performance in situations where specific technical skills and knowledge are required to perform the tasks.

Furthermore, there is evidence indicating causal relationships between stockperson attitudes and behaviour towards farm animals and the welfare and productivity of these animals. This research has been recently reviewed by Hemsworth and Coleman (2011) and Coleman and Hemsworth (2014) and, while most of this research has been conducted on farm animals in farm settings, this research has implications for handling and caring for farm animals post-farm gate.

Understanding the stockperson-sheep interactions at abattoirs and their outcomes for sheep welfare is necessary to identify, firstly, the stockperson attributes related to sheep welfare and thus, secondly, the training opportunities to target the key stockperson attributes necessary to safeguard sheep welfare at abattoirs. The importance of understanding human–animal interactions and their outcomes for the welfare and productivity of the animal has been clearly demonstrated in on-farm research in the livestock industries in which training programmes targeting the key attitudes and behaviour have been successfully introduced (Hemsworth and Coleman, 2011; Coleman and Hemsworth, 2014)): cognitive-behavioural techniques basically involve changing a person's behaviour by first targeting both the beliefs that underlie the behaviour (attitude) and the behaviour in question, and second, maintaining these changed beliefs and behaviours. This process of inducing behavioural change is a comprehensive procedure in which all of the personal and external factors that are relevant to the behavioural situation are explicitly targeted. This includes addressing common perceived barriers to change, addressing defensiveness about previous behaviour, changing habits, providing follow-ups to reinforce changes as well as changing the relevant attitudes and behaviour.

This approach to training has been shown to be practical and effective among a wide range of stockpersons working in a variety of on-farm situations, providing strong evidence for introducing this type of training into the livestock industries (Hemsworth and Coleman, 2011; Coleman and Hemsworth, 2014). The results of limited research on stockperson-sheep interactions at abattoirs highlight the importance of comprehensively understanding the stockperson attributes related to sheep welfare since such an understanding provides the basis to develop a cognitive-behavioural training programme for abattoir stockpeople to target those stockperson attitudes and behaviour that are related to sheep welfare. Clearly further research is required to comprehensively understand these key stockperson characteristics that affect sheep welfare at abattoirs.

Published by Burleigh Dodds Science Publishing Limited, 2021.

Further research on the influence of handling during loading and unloading sheep during road and sea transport is also required to fully understand the limiting stockperson characteristics and thus where training opportunities are required.

8.3 Welfare monitoring in the field

There are official (governmental or intergovernmental) animal welfare standards and private standards developed in the private sector, for example, by food industry stakeholders or non-governmental organizations (Matthews and Hemsworth, 2012). Private standards that have been created by large meat-buying customers, legislated standards or the World Organization for Animal Health (OIE) standards are used by private industry and some governments to assess animal welfare at abattoirs (Grandin, 2010).

The assessment of welfare of sheep post-farm gate can be used to demonstrate compliance with policy, law and regulatory standards, and to assure both consumers and non-consumers that certain welfare standards are being met. In recent years, the assessment and monitoring of animal welfare has shifted from the conventional approach of evaluating the environment and resources required to ensure good welfare, and instead has focused on animal-based measures of welfare. An example of this is the animal welfare assessment at a farm or on-site level in the European Union Welfare Quality® project (Botreau et al., 2009).

Animal-based measures can provide a direct assessment of the animal's welfare state, and while environmental parameters will offer information regarding potential or current welfare risks, they fail to directly reflect the welfare state of the animal (Colditz et al., 2014). Nevertheless, Main et al. (2014) suggest that outcome measures are unlikely to replace all environment measures, particularly where welfare science has shown that the resources provided contribute to genuine welfare benefits.

A number of authors have proposed a continuous improvement approach utilising regular monitoring of pre-defined welfare criteria (input (environmental and management) and outcome-based), benchmarking performance to identify targets for improvement and a management system to ensure preventive and corrective action to maximise levels of these criteria (Butterworth et al., 2011; Colditz et al., 2014; Main et al., 2014). Further, the authors suggest that such an animal welfare risk assessment and management scheme lends itself to providing evidence for compliance and assurance schemes. Thus this multi-pronged approach provides opportunities to benchmarking sheep welfare at each step post-farm gate to provide compliance evidence and market assurance and, probably most importantly, to utilize an animal welfare risk assessment and management scheme to continuously improve animal welfare.

Published by Burleigh Dodds Science Publishing Limited, 2021.

9 Future trends and conclusion

On-going improvements in sheep welfare in the future are likely to occur through research and adoption in the areas of animal management and facility design. Transport is a stressful procedure, especially when combined with other procedures, including mustering, yarding, handling, mixing and loading. The effects of land transport on physiological stress and immunological responses generally diminish after several hours, so that only the effects of food and water deprivation remain apparent. The key welfare issues for road transport are floor space allowance, journey duration and environmental conditions, such as THI, road conditions, and vehicle motion and noise.

The influence of handling during loading and unloading sheep during road and sea transport has not been comprehensively studied. However, based on research on animals on farms (Hemsworth and Coleman, 2011) and limited research at abattoirs (see Sections 4 and 6), further research on the effects of loading and unloading on transported sheep is warranted.

As with land transport, sea transport is stressful, involving a number of procedures in addition to transport *per se* such as handling during loading/unloading, loading density and environmental conditions. Furthermore, because of the extended period of transport there is increased risk of cumulative stress. While mortality rates have declined over the last few decades, key welfare issues are inanition and salmonellosis, space allowance, and environmental conditions. Research on the effects of space allowances on sheep behaviour, stress physiology, morbidity and mortality under different thermal conditions and different conditions of sea transport is required.

At abattoirs from the time of unloading to the time of slaughter, sheep experience a number of stressors. Well-designed facilities and properly trained stockpeople using judicious use of interaction with sheep as well as judicious use of dogs can minimise negative interactions when handling, and therefore minimise stress in sheep prior to slaughter. Disturbance and overcrowding in lairage should be avoided and sheep should have access to water at all times while in lairage. While restraint during stunning may be stressful in itself, proper restraint assists with the proper placement of the captive bolt or electrodes, ensuring a more effective stun and an animal that is insensible during neck incision.

Irrespective of the design of the transport and holding system, the skills, knowledge and attitudes of the stockpeople are integral to the standard of welfare experienced by the sheep. Attitudes influence not only the manner in which stockpeople handle sheep, but also their motivation to care for sheep. Thus, training targeting technical skills and knowledge as well as the attitudes and behaviours of stockpeople should be a primary component of the human resource management practices in all components of the supply chain.

Published by Burleigh Dodds Science Publishing Limited, 2021.

Welfare monitoring is an important tool to continuously improve animal welfare post-farm gate. It allows benchmarking at each of the steps post-farm gate, provides compliance evidence and market assurance and, probably most importantly, offers the opportunity to incorporate animal welfare risk assessment and management in the routine management of animals.

10 Where to look for further information

There are a number of reviews on animal welfare and its assessment (Fraser, 2008; Mellor et al., 2009; Mellor, 2012; Hemsworth et al., 2015). Fisher et al. (2009), EFSA (2011) and Cockram (2014) have reviewed the effects of road transport on sheep welfare and Phillips and Santurtun (2013) and Foster and Overall (2014) have reviewed the effects of sea transport on sheep welfare. The topic of human–animal interactions in the livestock industries has been reviewed by Hemsworth and Coleman (2011) and Coleman and Hemsworth (2014). Gregory (1998) has reviewed the relationships between animal welfare and meat quality.

11 Acknowledgements

The authors acknowledge G. Hutson and the late J. L. Barnett for their contributions with PHH to an earlier internal document on sheep welfare that was utilised in this review.

12 References

Anil, M. H. (2012), *Effects of Slaughter Method on Carcass and Meat Characteristics in the Meat of Cattle and Sheep*. EBLEX – a Division of the Agriculture and Horticulture Development Board, UK.

Anil, M. H. and McKinstry, J. L. (1991), 'Reflexes and loss of sensibility following head-to-back electrical stunning in sheep'. *Vet. Rec.*, 128, 106-7.

Anon. (1997), *The Welfare of Animals (Transport) Order 1997*. The Stationary Office Ltd., PO Box 276, London, SW8 5DT.

Anon (2002), *Model Code of Practice for the Welfare of Animals: Livestock at Slaughtering Establishments*. Standing Committee on Agriculture and Resource Management. CSIRO Publishing, Collingwood, Victoria, Australia.

Australian Government (2016), *Reports to Parliament,* http://www.agriculture.gov.au/export/controlled-goods/live-animals/live-animal-export-statistics/reports-to-parliament, accessed 8 May 2016.

Beausoleil, N. J., Stafford, K. J. and Mellor, D. J. (2005), 'Sheep show more aversion to a dog than to a human in an arena test'. *Appl. Anim. Behav. Sci.*, 91, 219-32.

Berg, C., Nordensten, C., Hultgren, J. and Algers, B. (2012), 'The effect of stun duration and level of applied current on stun and meat quality of electrically stunned lambs under commercial conditions'. *Anim. Welfare.*, 21(Suppl. 2), 131-8.

Black, H. (1996), 'Inanition, stress and immunity in the expression of salmonellosis in the live sheep export industry'. *New. Zeal. Vet. J.*, 44, 77-8.

Blackmore, D. K. and Petersen, G. V. (1981), 'Stunning and slaughter of sheep and calves in New Zealand'. *New. Zeal. Vet. J.*, 29, 99-102.

Boissy, A., Manteuffel, G., Jensen, M. B., Oppermann, M., Spruijt, B., Keeling, L. J., Winckler, C., Forkman, B., Dimitrov, I., Langbein, J., et al. (2007), 'Assessment of positive emotions in animals to improve their welfare'. *Physiol. Behav.*, 92, 375-97.

Botreau, R., Veissier, I. and Perny, P. (2009), 'Overall assessment of animal welfare: strategy adopted in Welfare Quality®'. *Anim. Welfare*, 18, 363-70.

Broom, D. M., Goode, J. A., Hall, S. J. G., Lloyd, D. M. and Parrott, R. F. (1996), 'Hormonal and physiological effects of a 15 hour road journey in sheep: comparison with the responses to loading, handling and penning in the absence of transport'. *Brit. Vet. J.*, 152, 593-604.

Butterworth, A., Mench, J. A. and Wielebnowski, N. (2011), 'Practical strategies to assess and improve welfare'. In M. C. Appleby, B. Hughes and J. Mench (Eds), *Animal Welfare*, CABI, Wallingford, Oxfordshire, UK, pp. 200-14.

Caulfield, M. P., Cambridge, H., Foster, S. F. and McGreevy, P. D. (2014), 'Heat stress: A major contributor to poor animal welfare associated with long-haul live export voyages'. *Vet. J.*, 199, 223-8.

Cockram, M. S. (2007), 'Criteria and potential reasons for maximum journey times for farm animals destined for slaughter'. *Appl. Anim. Behav. Sci.*, 106, 234-43.

Cockram, M. S. (2014), 'Sheep transport'. In T. Grandin (Ed.), *Livestock Handling and Transport*, 4th Edition, CAB International, Oxon, UK, pp. 228-4.

Cockram, M. S., Baxter, E. M., Smith, L. A., Bell, S., Howard, C. M., Prescott, R. J. and Mitchell, M. A. (2004), 'Effect of driver behaviour, driving events and road type on the stability and resting behaviour of sheep in transit'. *Anim. Sci.*, 79, 165-76.

Cockram, M. S., Kent, J. E., Goddard, P. J., Waran, N. K., McGilp, I. M., Jackson, R. E., Muwanga, G. M. and Prytherch, S. (1996), 'Effect of space allowance during transport on the behavioural and physiological responses of lambs during and after transport'. *Anim. Sci.*, 62, 461-77.

Cockram, M. S., Kent, J. E., Jackson, R. E., Goddard, P. J., Doherty, O. M., McGilp, I. M., Fox, A., Studdert-Kennedy, T. C., McConnell, T. I. and O'Riordan, T. (1999a), 'Effect of lairage during 24 h of transport on the behavioural and physiological responses of sheep'. *Anim. Sci.*, 65, 391-402.

Cockram, M. S., Kent, J. E., Waran, N. K., McGilp, I. M., Jackson, R. E., Amory, J. R., Southall, E. L., O'Riordan, T., McConnell, T. I. and Wilkins, B. S. (1999b), 'Effects of a 15h journey followed by either 12h starvation or ad libitum hay on the behaviour and blood chemistry of sheep'. *Anim. Welf.*, 8, 135-48.

Cockram, M. S., Kent, J. E., Goddard, P. J., Waran, N. K., Jackson, R. E., McGilp, I. M., Southall, E. L., Amory, J. R., McConnell, T. I., O'Riordan, T. and Wilkins, B. S. (2000), 'Behavioural and physiological responses of sheep to 16 h transport and a novel environment post-transport'. *Vet. J.*, 159, 139-46.

Colditz, I., Ferguson, D., Collins, T., Matthews, L. and Hemsworth, P. (2014), 'A prototype tool to enable farmers to measure and improve the welfare performance of the farm animal enterprise: The Unified Field Index'. *Animals*, 4, 446-62.

Coleman, G. J. (2010), 'Educating the public: information or persuasion?'. *J. Vet. Med. Ed.*, 37, 74-82.

Coleman, G. J., Rohlf, V., Toukhsati, S. and Blache, D. (2015), 'Public attitudes relevant to farm animal welfare policy'. *Farm Policy J.*, 12(3), 45–57.

Coleman, G. J., Rice, M. and Hemsworth, P. H. (2012), 'Human-animal relationships at sheep and cattle abattoirs'. *Anim. Welfare*, 21(S2), 15–21.

Coleman, G. J. and Hemsworth, P. H. (2014), 'Training to improve stockperson beliefs and behaviour towards livestock enhances) welfare and productivity'. *Scientific and Technical Review of the Office International des Epizooties (Paris)*, 33 (1), 131–7.

Deiss, V., Temple, D., Ligout, S., Racine, C., Bouix, J., Terlouw, C. and Boissy, A. (2009), 'Can emotional reactivity predict stress responses at slaughter in sheep'. *Appl. Anim. Behav. Sci.*, 119, 193–202.

De Jonge, J. and van Trijp, H. C. M. (2013), 'Meeting heterogeneity in consumer demand for animal welfare: A reflection on existing knowledge and implications for the meat sector'. *J. Agric. Environ. Ethics*, 26, 629–61.

Dwyer, C. M. and Lawrence, A. B. (2008), 'Introduction to animal welfare and the sheep'. In C. M. Dwyer (Ed.), *The Welfare of Sheep*, Springer, New York, pp. 1–40.

EFSA (2004), 'Opinion of the Scientific Panel on Animal Health and Welfare on a request from the Commission related to welfare aspects of the main systems of stunning and killing the main commercial species of animals'. *EFSA J.*, 45, 1–29.

EFSA (2006), 'Opinion of the Scientific Panel on Animal Health and Welfare on a request from the Commission related with the welfare aspects of the main systems of stunning and killing applied to commercially farmed deer, goats, rabbits, ostriches, ducks, geese and quail'. *EFSA J.*, 326, 1–18.

EFSA (2011), 'Scientific opinion concerning the welfare of animals during transport'. [125 pp.]. *EFSA J.*, 9(1), 1966.

Eldridge, G. A., Warner, R. D., Winfield, C. G. and Vowles, W. J. (1989), 'Pre-slaughter management and marketing systems for cattle in relation to improving meat yield, meat quality and animal welfare'. *Report for Australian Meat and Livestock Research and Development Corporation*, Werribee, Victoria, Australia.

Farouk, M. M., Al-Mazeedi, H. M., Sabow, A. B., Bekhit, A. E. D., Adeyemi, K. D., Sazili, A. Q. and Ghani, A. (2014), 'Halal and Kosher slaughter methods and meat quality: A review'. *Meat Sci.*, 98, 505–19.

Ferguson, D. M. and Lea, D. (2013), 'Refining stocking densities'. *Final Report W.LIV.0253 for Meat and Livestock Australia Limited*, North Sydney NSW Australia.

Ferguson, D. M., Schreurs, N. M., Kenyon, P. R. and Jacob, R. H. (2014), 'Balancing consumer and societal requirements for sheep meat production: An Australasian perspective'. *Meat Sci.*, 98, 477–83.

Ferguson, D. M. and Warner, R. D., (2008), 'Have we underestimated the effect of pre-slaughter stress on meat quality in ruminants?'. *Meat Sci.*, 80, 12–19.

Fisher, A. D. (2013), 'Can live animal export ever be humane?' *Issues*, June 2013, 103. http://www.issuesmagazine.com.au/article/issue-june-2013/can-live-animal-export-ever-be-humane.html.

Fisher, A. D., Niemeyer, D. O., Lea, J. M., Lee, C., Paull, D. R., Reed, M. T. and Ferguson, D. M. (2010), 'The effects of 12, 30, or 48 hours of road transport on the physiological and behavioral responses of sheep'. *J. Anim. Sci.*, 88, 2144–52.

Fisher, A. D., Colditz, I. G., Lee, C. and Ferguson, D. M. (2009), 'The influence of land transport on animal welfare in extensive farming systems'. *J. Vet. Behav. Clin. Appl. Res.*, 4, 157–62.

Fisher, A. D., Stewart, M., Duganzich, D. M., Tacon, J. and Matthews, L. R. (2005), 'The effects of stationary periods and external temperature and humidity on thermal stress conditions within sheep transport vehicles'. *New Zeal. Vet. J.*, 53, 6-9.

Fisher, M. W. and Mellor, D. J. (2008), 'Developing a systematic strategy incorporating ethical, animal welfare and practical principles to guide the genetic improvement of dairy cattle'. *New Zeal Vet. J.*, 65, 100-6.

Foster, S. F. and Overall, K. L. (2014), 'The welfare of Australian livestock transported by sea'. *Vet J.*, 200, 205-9.

Fraser, D. (2008), *Understanding Animal Welfare: The Science in its Cultural Context*, Wiley-Blackwell, West Sussex, United Kingdom.

Gibson, T. J., Johnson, C. B., Murrell, J. C., Hulls, C. M., Mitchinson, S. L., Stafford, K. J., Johnstone, A. C. and Mellor, D. J. (2009a), 'Electroencephalographic responses of halothane-anaesthetised calves to slaughter by ventral-neck incision without prior stunning'. *New Zeal. Vet. J.*, 57, 77-83.

Gibson, T. J., Johnson, C. B., Murrell, J. C., Chambers, J. P., Stafford, K. J. and Mellor, D. J. (2009b), 'Components of electroencephalographic responses to slaughter in halothane-anaesthetised calves: Effects of cutting neck tissues compared with major blood vessels'. *New Zeal. Vet. J.*, 57, 84-9.

Gonyou, HW, Stookey, JM and McNeal, LG. (1985), 'Effects of double decking and space allowance on the performance and behaviour of feeder lambs'. *J. Anim. Sci.*, 60, 1110-16.

Grandin, T. (1997), 'Assessment of stress during handling and transport'. *J. Anim. Sci.*, 75, 249-57.

Grandin, T. (2007), 'Introduction: Effect of customer requirements, international standards and marketing structure on the handling and transport of livestock and poultry'. In T. Grandin (Ed.), *Livestock Handling and Transport*, CAB International, Oxon, UK, pp. 1-18.

Grandin, T. (2010). 'Auditing animal welfare at slaughter plants'. *Meat Sci.*, 86, 56-65.

Gray J. A. (1987), *The Psychology of Fear and Stress*, 2nd Edition, Cambridge University Press, Cambridge, Massachusetts, USA.

Gregory, N. G. (1998), *Animal Welfare and Meat Science*, 2nd Edition, CAB International, Oxon, United Kingdom.

Gregory, N. G., Fielding, H. R., von Wenzlawowicz, M. and von Holleben, K. (2010), 'Time to collapse following slaughter without stunning in cattle'. *Meat Sci.*, 85, 66-9.

Gregory, N. G. and Wotton, S. B. (1984). 'Sheep slaughtering procedures. III. Head-to-back electrical stunning'. *Br. Vet. J.*, 140, 570-5.

Hall, S. J. G., Schmidt, B. and Broom, D. M. (1997), 'Feeding behaviour and the intake of food and water by sheep after a period of deprivation lasting 14 h'. *Anim. Sci.*, 64, 105-10.

Hall, S. J. G., Kirkpatrick, S. M., Lloyd, D. M. and Broom, D. M. (1998), 'Noise and vehicular motion as potential stressors during the transport of sheep'. *Anim. Sci.*, 67, 467-73.

Hall, S. J. G. and Bradshaw, R. H. (1998), 'Welfare aspects of the transport by road of sheep and pigs'. *J. Appl. Anim. Welfare Sci.*, 1, 235-54.

Hargreaves, A. L. and Hutson, G. D. (1997), 'Handling systems for sheep'. *Livest. Prod. Sci.*, 49, 121-38.

Hemsworth, P. H. and Coleman, G. J. (2009), 'Animal welfare and management'. In F. J. M. Smulders and B. Algers (Eds), *Food Safety Assurance and Veterinary Public*

Health Volume 5. Welfare Productions Animals: Assessment and Management Risks, Wageningen Academic Publishers, the Netherlands, pp. 133–47.

Hemsworth, P. H. and Coleman, G. J. (2011), Human-Livestock Interactions: The Stockperson and the Productivity and Welfare of Farmed Animals, 2nd Edition CAB International, Oxon United Kingdom.

Hemsworth, P. H., Mellor, D. J., Cronin, G. M. and Tilbrook, A. J. (2015), 'Scientific assessment of animal welfare', New Zeal. Vet. J., 63, 24–30.

Hemsworth, P. H., Rice, M., Karlen, M. G., Calleja, L., Barnett, J. L., Nash, J. and Coleman, G. J. (2011), 'Human-animal interactions at abattoirs: Relationships between handling and animal stress in sheep and cattle'. Appl. Anim. Behav. Sci., 135, 24–33.

Hemsworth, P. H., Rice, M., Borg, S., Edwards, L. E., Ponnampalam, E. N. and Coleman, G. J. (2016), 'Relationships between handling, behaviour and stress in lambs at abattoirs'. In Proc. 31st Biennial Conf. of the Aust. Soc. of Animal Prod., p. 79.

Higgs, A. R. B., Norris, R. T. and Richards, R. B. (1991), 'Season, age and adiposity influence death rates in sheep exported by sea'. Aust. J. Agric. Res., 42, 205–14.

Higgs, A. R. B., Norris, R. T. and Richards, R. B. (1993), 'Epidemiology of salmonellosis in the live sheep export industry'. Aust. Vet. J., 70, 330–5.

Higgs, A. R. B., Norris, R. T., Love, R. A. and Norman, G. J. (1999), 'Mortality of sheep exported by sea: evidence of similarity by farm group and of regional differences'. Aust. Vet. J., 77, 729–33.

Hodge, R. W., Watson, M. J., Butler, K. L., Kelly, A. J., Beers, P. J. and Bogdanovic, B. W. (1991), 'Export of live sheep: nutritional studies and the failure to eat syndrome'. Rec. Adv. An. Aust., 394 113–20.

Hogan, J. P., Petherick, J. C. and Phillips, C. J. C. (2007), 'The physiological and metabolic impacts on sheep and cattle of feed and water deprivation before and during transport'. Nutr. Res. Rev., 20, 17–28.

Hogan, J. A. (2008), 'Motivation'. In J. J. Bolhuis and L.-C. Giraldeau (Eds), The Behaviour of Animals: Mechanisms, Function and Evolution, Blackwell Publishing, Malden, Massachusetts, pp. 41–70.

Hutson, G. D. (1982), 'Flight distance' in Merino sheep.' Anim. Prod., 35, 231–5.

Jackson, R. E., Cockram, M. S., Goddard, P. J., Doherty, O. M., McGilp, I. M. and Waran, N. K. (1999), 'The effects of 24 h water deprivation when associated with some aspects of transportation on the behaviour and blood chemistry of sheep'. Anim. Welfare., 8, 229–41.

Jacob, R. H., Pethick, D. W. and Chapman, H. M. (2005), 'Muscle glycogen concentrations in commercial consignments of Australian lamb measured on farm and post-slaughter after three different lairage periods'. Anim. Prod. Sci., 45, 543–52.

Jacob, R. H., Pethick, D. W., Ponnampalam, E., Speijers, J. and Hopkins, D. L. (2006), 'The hydration status of lambs after lairage at two Australian abattoirs'. Anim Prod. Sci., 46, 909–12.

Jarvis, A. M. and Cockram, M. S. (1995), 'Some factors affecting resting behaviour of sheep in slaughterhouse lairages after transport from farms'. Anim. Welfare, 4, 53–60.

Johnson, C. B., Gibson, T. J., Stafford, K. J. and Mellor, D. J. (2012), 'Pain perception at slaughter'. Anim. Welfare, 21 (Supp. 2), 113–22.

Johnson, C. B., Mellor, D. J., Hemsworth, P. H. and Fisher, A. D. (2015), 'A scientific comment on the welfare of domesticated ruminants slaughtered without stunning'. NZ Vet. J., 63, 58–65.

Jolly, S. and Wallace, A. (2007), *Best Practice for Production Feeding of Lambs: A Review of the Literature*, Meat and Livestock Australia, Sydney, Australia.

Jones, T. A., Waitt, C. and Dawkins, M. S. (2010), 'Sheep lose their balance, slip and fall less when loosely packed in transit where they stand close to but not touching their neighbours'. *Appl. Anim. Behav. Sci.*, 123, 16-23.

Jongman, E. C., Rice, M., Campbell, A. J. D., Butler, K. L. and Hemsworth, P. H. (2016), 'The effect of trough space and floor space on feeding and welfare of lambs in an intensive finishing system'. *Appl. Anim. Behav. Sci.* (In press).

Jongman, E. C., Edge, M. K., Butler, K. L. and Cronin, G. M., (2008), 'Reduced space allowance for adult sheep in lairage for 24 hours limits lying behaviour but not drinking behaviour'. *Anim. Prod. Sci.*, 48, 1048-51.

Kelly, A. P. (1990), *Health and Welfare Research in the Live Sheep Export Trade*, Department of Agriculture and Rural Affairs, Hamilton, Victoria, Australia.

Kelly, A. P. (1995), *Mortalities in Sheep Transported by Sea*. PhD Thesis, Faculty of Veterinary Science, University of Melbourne, Australia.

Kettlewell, P. J., Hoxey, R. P., Hartshorn, R. L., Meeks, I. R. and Twydell, P. (2001), 'Controlled ventilation system for livestock transport vehicles'. In R. R. Stowell, R. Bucklin, and R. W. Bottcher (Eds), *Livestock Environment VI, Proceedings of the Sixth International Symposium*, American Society of Agricultural Engineers, St. Joseph, Michigan, pp. 556-63.

Kilgour, R. and de Langen, H., (1970), 'Stress in sheep resulting from management practices'. *Proc. New. Zeal. Soc. of Anim. Prod.*, 30, 65.

Kim, F. B., Jackson, R. E., Gordon, G. D. H. and Cockram, M. S. (1994), 'Resting behaviour of sheep in a slaughterhouse lairage'. *Appl. Anim. Behav. Sci.*, 40, 45-54.

Kirby, R. M., Jones, F. M., Ferguson, D. M. and Fisher, A. D. (2004), 'Adaptation to grain feeding'. *Feeding Grain for Sheep Meat Production*, Sheep Co-operative Research Corporation, Armidale, NSW, Australia.

Kjaernes, U. and Lavik, R. (2008), 'Opinions on animal welfare and food consumption in seven European countries'. In U. Kjaernes, B. B. Bock, E. Roe and J. Roex (Eds), *Consumption, Distribution And Production Of Farm Animal Welfare – Opinions And Practices Within The Supply Chain*, Welfare Quality® Reports No. 7, Cardiff University, Cardiff, UK, pp. 1-126.

Knowles, T. G. (1998), 'A review of the road transport of slaughter sheep'. *Vet. Rec.*, 143, 212-19.

Knowles, T. G., Maunder, D. H. L. and Warriss, P. D. (1994), 'Factors affecting mortality of lambs in transit or in lairage at a slaughterhouse, and reasons for carcase condemnations'. *Vet. Rec.*, 135, 109-11.

Knowles, T. G., Brown, S. N., Warriss, P. D., Phillips, A. J., Dolan, S. K., Hunt, P., Ford, J. E., Edwards, J. E. and Watkins, P. E. (1995), 'Effects on sheep of transport by road for up to 24 hours'. *Vet. Rec.*, 136, 431-8.

Knowles, T. G., Warriss, P. D., Brown, S. N., Kestin, S. C., Edwards, J. E., Perry, A. M., Watkins, P. E. and Phillips, A. J. (1996), 'Effects of feeding, watering and resting intervals on lambs transported by road and ferry to France'. *Vet. Rec.*, 139, 335-9.

Knowles, T. G., Warriss, P. D., Brown, S. N. and Edwards, J. E. (1998), 'Effects of stocking density on lambs being transported by road'. *Vet. Rec.*, 142, 503-9.

Krawczel, P. D., Friend, T. H., Caldwell, D. J., Archer, G. and Ameiss, K. (2007), 'Effects of continuous versus intermittent transport on plasma constituents and antibody response of lambs'. *J. Anim. Sci.*, 85, 468-76.

Published by Burleigh Dodds Science Publishing Limited, 2021.

Launchbaugh, K. L., Provenza, F. D. and Werkmeister, M. J. (1997), 'Overcoming food neophobia in domestic ruminants through addition of a familiar flavor and repeated exposure to novel foods'. *Appl. Anim. Behav. Sci.*, 54, 327-34.

LiveCorp (2014), Sheep Statistics, LiveCorp, Australian Livestock Export Corporation Ltd., North Sydney, NSW, Australia, Available: http://www.livecorp.com.au/ industry-information/industry-statistics/sheep-statistics.

Lynch, J. J. (1988), 'Behaviour and the export of live sheep'. In B. Farquharson, J. Lynch and R. Kellaway (Eds), *Standing Committee on Agriculture Workshop on Livestock Export Research*, Bureau of Rural Resources, Department of Primary Industries and Energy, Proceedings No. 3, Australian Government Publishing Service, Canberra, pp. 155-66.

Lynch, J. J. and Bell, A. K. (1987), 'The transmission from generation to generation in sheep of the learned behaviour for eating grain supplements'. *Aust. Vet. J.*, 64, 291-2.

Main, D., Mullan, S., Atkinson, C., Cooper, M., Wrathall, J. and Blokhuis, H. (2014), 'Best practice framework for animal welfare certification schemes'. *Trends Food Sci. Tech.*, 37, 127-36.

Matthews, L. R. and Hemsworth, P. H. (2012), 'Drivers of change: law, international markets, and policy'. *Anim. Front.*, 2, 40-5.

Mellor DJ. (2012), 'Animal emotions, behaviour and the promotion of positive welfare states'. *New Zeal. Vet. J.*, 60, 1-8.

Mellor, D. J., Patterson-Kane, E. and Stafford, K. J. (2009), *The Sciences of Animal Welfare*, Wiley-Blackwell Publishing, Oxford, United Kingdom.

Mellor, D. J. and Bayvel, A. C. D. (2008), 'New Zealand's inclusive science-based system for setting animal welfare standards'. *Appl. Anim. Behav. Sci.*, 113, 313-29.

Miranda-de la Lama, G. C., Monge, P., Villarroel, M., Olleta, J. L., García-Belenguer, S. and María, G. A. (2011), 'Effects of road type during transport on lamb welfare and meat quality in dry hot climates'. *Trop. Anim. Health Prod.*, 43, 915-22.

Nakyinsige, K., Man, Y. C., Aghwan, Z. A., Zulkifli, I., Goh, Y. M., Bakar, F. A., Al-Kahtani, H. A. and Sazili, A. Q. (2013), 'Stunning and animal welfare from Islamic and scientific perspectives'. *Meat Sci.*, 95, 352-61.

Njisane, Y. Z. and V. Muchenje (2013), 'Quantifying avoidance-related behaviour and bleeding times of sheep of different ages, sex and breeds slaughtered at a municipal and a commercial abattoirs'. *S. Afr. J. Anim. Sci.*, 43, 38-42.

Norris, R. T. (2005), 'Transport of animals by sea'. *Revue Scientifique Et Technique* (Office International Des Epizooties), 24, 673-81.

Norris, R. T. and Richards, R. B. (1989), 'Deaths in sheep exported by sea from Western Australia - analysis of ship Master's reports'. *Aust. Vet. J.*, 66, 97-102.

Norris, R. T., McDonald, C. L., Richards, R. B., Hyder, M. W., Gittins, S. P. and Norman, G. J. (1990a), 'Management of inappetant sheep during export by sea'. *Aust. Vet. J.*, 67, 244-7.

Norris, R. T., Richards, R. B. and Dunlop, R. H. (1990b), 'Pre-embarkation risk factors for sheep deaths during export by sea from Western Australia'. *Aust. Vet. J.*, 66, 309-14.

Norris, R. T., Richards, R. B. and Higgs, A. R. (1990c), 'Research on the health, husbandry and welfare of sheep during live export'. *J. Agric. Western Australia*, 31, 131-48.

Norris, R. T., Richards, R. B. and Norman, G. J. (1992), 'The duration of lot-feeding of sheep before sea transport'. *Aust. Vet. J.*, 69, 8-10.

OIE (World Organisation for Animal Health) (2010), *Roles of public and private standards in animal health and animal welfare*. Accessed 27 May 2016. http://www.oie.int/fileadmin/Home/eng/Internationa_Standard_Setting/docs/pdf/A_RESO_2010_PS.pdf.

Parrott, R. F., Hall, S. J. G. and Lloyd, D. M. (1998a), 'Heart rate and stress hormone responses of sheep to road transport following two different loading procedures'. *Anim. Welf.,* 7, 257–67.

Parrott, R. F., Hall, S. J. G., Lloyd, D. M., Goode, J. A. and Broom, D. M. (1998b), 'Effects of a maximum permissible journey time (31 h) on physiological responses of fleeced and shorn sheep to transport, with observations on behaviour during a short (1 h) rest-stop'. *Anim. Sci.,* 66, 197–207.

Parrott, R. F., Lloyd, D. M. and Goode, J. A. (1996), 'Stress hormone responses of sheep to food and water deprivation at high and low ambient temperatures'. *Anim. Welf.,* 5, 45–56.

Petherick, J. C. (2007), 'Spatial requirements of animals: allometry and beyond'. *J. Vet. Behav.,* 2, 197–204.

Petherick, J. C. and Phillips, C. J. C. (2009), 'Space allowances for confined livestock and their determination from allometric principles'. *Appl. Anim. Behav. Sci.,* 117, 1–12.

Phillips, C. J. C. (2005), 'Ethical perspectives of the Australian live export trade'. *Aust. Vet. J.,* 83, 558–62.

Phillips, C. J. C. (2008), 'The welfare of livestock during sea transport'. In M. C. Appleby, V. A. Cussen, L. Garcés, L. A. Lambert and J. Turner (Eds), *Long Distance Transport and Welfare of Farm Animals,* CAB International, Oxon United Kingdom, pp. 137–54.

Phillips, C. J. C. and Santurtun, E. (2013), 'The welfare of livestock transported by ship'. *Vet J.,* 196, 309–14.

Pines, M. K. and Phillips, C. J. C. (2011), 'Accumulation of ammonia and other potentially noxious gases on live export shipments from Australia to the Middle East'. *J. Envir. Monitor.,* 13, 2798–807.

Phillips, C. J. C., Pines, M. K., Latter, M., Muller, T., Petherick, J. C., Norman, S. T. and Gaughan, J. B. (2012a), 'The physiological and behavioral responses of sheep to gaseous ammonia'. *J. Anim. Sci.,* 90, 1562–9.

Phillips, C. J., Pines, M. K. and Muller, T. (2012b), 'The avoidance of ammonia by sheep'. *J. Vet. Behav. Clinic. Applic. Res.,* 7, 43–8.

Randall, J. M. (1993), 'Environmental parameters necessary to define comfort for pigs, cattle and sheep in livestock transporters'. *Anim. Prod.,* 57, 299–307.

Richards, R. B., Norris, R. T., Dunlop, R. H. and McQuade, N. C. (1989), 'Causes of death in sheep exported live by sea'. *Aust. Vet. J.,* 66, 33–8.

Richards, R. B., Hyder, M. W., Fry, J., Costa, N. D., Norris, R. T. and Higgs, A. R. B. (1991), 'Seasonal metabolic factors may be responsible for deaths in sheep exported by sea'. *Aust. J. Agric. Res.,* 42, 216–26.

Ruiz-de-la-Torre, J. L., Velarde, A., Diestre, A., Gispert, M., Hall, S. J. G., Broom, D. M. and Manteca, X. (2001), 'Effects of vehicle movements during transport on the stress responses and meat quality of sheep'. *Vet. Rec.,* 148, 227–9.

Rushen, J. (1986), 'Aversion of sheep for handling treatments: paired-choice studies'. *Appl. Anim. Behav. Sci.,* 16, 363–70.

Savage, D. B., Ferguson, D. M., Fisher, A. D., Hinch, G. N., Mayer, D. G., Duflou, E., Lea, J. M., Baillie, N. D. and Raue, M., (2008), 'Preweaning feed exposure and different feed delivery systems to enhance feed acceptance of sheep'. *Anim. Prod. Sci.,* 48, 1040–3.

Shorthose, W. R. and Wythes, J. R. (1988), 'Transport of sheep and cattle'. In *Proceedings 34th Congress of Meat Science and Technology*, Brisbane, Australia, Part A, pp.122-9.

Stockman, C. A., Barnes, A. L., Maloney, S. K., Taylor, E., McCarthy, M. and Pethick, D. (2011), 'Effect of prolonged exposure to continuous heat and humidity similar to long haul live export voyages in Merino wethers'. *Anim. Prod. Sci.*, 51, 135-43.

Syme, L. A. (1985), *Intensive Sheep Management, with Particular Reference to the Live Sheep Trade*, Australian Agricultural Health and Quarantine Service, Department of Primary Industry, Australian Government Publishing Service, Canberra.

Terlouw, E. M. C., Arnould, C., Auperin, B., Berri, C., Le Bihan-Duval, E., Deiss, V., Lefevre, F., Lensink, B. J. and Mounier, L., (2008), 'Pre-slaughter conditions, animal stress and welfare: current status and possible future research'. *Animal*, 2, 1501-17.

Thwaites, C. J. (1985), 'Physiological responses and productivity in sheep'. In M. K. Yousef (Ed.), *Stress Physiology in Livestock, Volume II Ungulates*, CRC Press, USA, pp. 25-38.

Toohey, E. S. and Hopkins, D. L. (2006), 'Effects of lairage time and electrical stimulation on sheep meat quality'. *Anim. Prod. Sci.*, 46, 863-7.

Turner, S. and Dwyer, C. (2007), 'Welfare assessment in extensive animal production systems: challenges and opportunities'. *Animal Welfare*, 16, 189-92.

Vanhonacker, F., Verbeke, W., Van Poucke, E., Pieniak, Z., Nijs, G. and Tuyttens, F. A. M. (2012), 'The concept of farm animal welfare: citizen perceptions and stakeholder opinion in Flanders'. *Belgium. J. Agric. Environ. Ethics*, 25, 79-101.

Vanhonacker, F. and Verbeke, W. (2014), 'Public and consumer policies for higher welfare food products: challenges and opportunities'. *J. Agric. Environ. Ethics*, 27, 153-71.

Velarde, A., Ruiz-de-la-Torre, J. L., Roselló, C., Fàbrega, E., Diestre, A. and Manteca, X. (2002), 'Assessment of return to consciousness after electrical stunning in lambs'. *Anim. Welfare*, 11, 333-41.

Velarde, A., Ruiz-de-la-Torre, J. L., Stub, C., Diestre, A. and Manteca, X., (2000), 'Factors affecting the effectiveness of head-only electrical stunning in sheep'. *Vet Rec.*, 147, 40-3.

Weeks, C. A. (2008), 'A review of welfare in cattle, sheep and pig lairages, with emphasis on stocking rates, ventilation and noise.' *Anim. Welfare.*, 17, 275-84.

CPSIA information can be obtained
at www.ICGtesting.com
Printed in the USA
BVHW010350231121
622254BV00011B/253

9 781801 462235